Projective Geome

Lawrence Edwards

Projective Geometry

Floris Books

First published in 1985 by Rudolf Steiner Institute, Inc. Pennsylvania
Second revised edition published by Floris Books in 2003
Third printing 2015
© 1985, 2003 Lawrence Edwards

British Library CIP Data available
ISBN 978-086315-393-8

Printed & Bound by MBM Print SCS Ltd, Glasgow

Contents

Foreword

Over twenty years ago a number of my friends in Edinburgh formed a small group for the study of Projective Geometry. We met together every fortnight over the course of several years and this book represents the substance of what I gave them then. Most of them were not trained mathematicians and the course was never intended to be a rigorously logical approach, but we wanted to see how far we could go in a clear and *consequential* manner, so that the mind and imagination would be able to perceive as far as possible just how, and why, one thing follows from another, without ascending (or descending?) into abstruseness. To this end the work is almost entirely synthetic; that is, open to the visual imagination, but the algebraic core, which, in the final analysis, must be seen as imparting to the subject its essential symmetries and beauty, has been recognized, and lightly touched upon in Chapter 5.

Much of the material in Chapter 13 and onwards, dealing with the imaginary points and lines, the path curves, and the intricacies of three-dimensional space, appears, as far as I know, nowhere else in English, in the form in which I have given it here; and I therefore dare to hope that the book will prove to be more than a re-presentation of facts which are already well-known.

At this point it is worth mentioning that the form which this book has taken is due to the author's conviction, supported by many years of teaching, that geometry is a subject which must be *done* as well as just being understood. Until it has entered into, and been experienced in the will, it has not been fully grasped. Even when one believes that one has fully understood a process in the intellect, it is still worth while actually doing the drawings and constructions, and it is remarkable how often in this process, new insights awaken. Therefore, many of the most beautiful and important things in the book are hidden in the exercises, and, especially in the later chapters, the instructions for model making. The exercises must be looked upon as containing buried treasure. The exercises should be reworked by the reader; endless variation is possible, and in exploring these possibilities one's aesthetic-geometric sense is greatly developed.

It is only right that I should record here my thanks to John and Norma Blackwood of Sydney without whose hard work and enthusiastic encouragement this book would surely never even have approached the possibility of publication. Also, I would like to thank Martin Levin of Toronto for his painstaking reading of the manuscript, and for the many pieces of good advice which he gave, especially at the points where otherwise the text might have seemed obscure.

For this edition I must thank Lou de Boer for his meticulous revision, as well as his work in scanning the text and diagrams from the earlier edition. My former colleague, Denis Wight, has worked through the proofs, checking many of the exercises, and catching a few errors which may have thrown the unsuspecting reader off the course.

Without all their work and effort this book could not have been published. May it prove worthy of their efforts.

Strontian, July 2003

1. Introduction

This course of lessons is not intended to be a closely reasoned system in which all the logical connections are rigorously treated, but is an attempt to lead to as clear and artistic an understanding of the qualities of space as can be attained without abstruseness. Where the words 'it will be found ...' or 'it can be shown ...' appear they mean that the following truth is one, the proof of which is beyond the bounds of this work. At some later date the student will doubtless wish to obtain the proof from one of the standard published works; meanwhile it is fruitful if he takes the truth empirically; that is, as a matter of observation. We do not always need to prove that the sun rises each morning! In such cases it will be valuable for the student to make several figures, and assure himself that, at any rate in the case of these, the stated truth holds good.

1.1 The elements of geometry

The fundamental objects one deals with in geometry, are Space, planes, lines, points and the Nothing. They are called the *elements* of geometry. In modern mathematics there is generally no attempt made to define them. Yet it is well worth reflecting a while on what these things are, and are like, before we begin.

Considering a *line*, we must distinguish from a stripe, though we use stripes as thin as possible to visualize lines. So the stripe through C and D in Figure 1.1 is 'better' than the one through A and B. The best one, through E and F has the disadvantage of being invisible. Yet there is another way to see our invisible lines.

Look at Figure 1.2: the line is the border between black and white. Note that every point in the figure is either black or white, so the line between them has no material existence.

A second property of the line is its extent: it has no borders; in neither of its both directions it has an endpoint, it has infinite length. By 'line' we always mean '*straight* line'; with 'straightness' we mean the shape of a beam of light, or of a stretched wire.

Notice that in order to introduce the line, we need a plane, Figure 1.2. We will meet other properties of the line that require a plane to

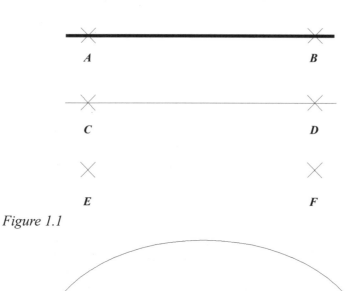

Figure 1.1

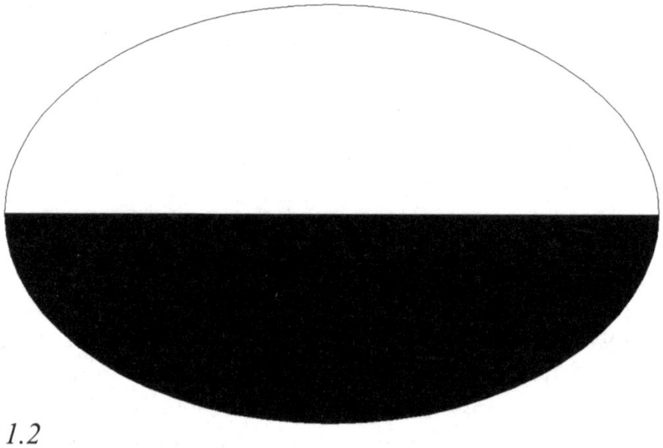

Figure 1.2

prove them. That suggests that the line can hardly exist without a second dimension.

If two lines in space meet, there come into existence both a meeting *point* and a joining *plane*. The point has no size at all; it merely determines a location. The plane is *extended* infinitely in all its directions, which are many, and it is *thin* and *flat*. Think of a lake in a calm afternoon, no waves at all: the border between water and air is a plane, it is infinitely thin. Flat means: containing (straight) lines in all directions of the plane.

Sometimes we will need two more elements. The *Space* — of which there is only one — is considered to be the biggest element, containing all other elements. At the other hand there is the *Nothing* (the Void, the Empty) — of which there is only one too — that acts as the smallest element, being present in all other elements. (If you add Nothing to a point, it won't change much, will it? So you must agree that a point contains also Nothing.)

1.2 Nomenclature

It is usual to denote points by capital letters, lines by small letters and planes by Greek letters. Thus A would be a point, b would be a line and α would be a plane. AB would be a line — that passing through the points A and B.

In plane geometry ab would be a point — the meeting point of lines a and b. AB,CD would again be a point — the meeting point of lines AB and CD. Conversely ab,cd would be a line — that passing through the points ab and cd.

In space, $\alpha\beta$ would be a line — that in which the planes α and β meet, etc.

A set of points on a line is called a *range* of points, and the set of lines lying in one point and in one plane is called a *pencil* of lines.

If three (or more) points are on one line, they are called *collinear*. If three (or more) lines pass through one point, they are called *concurrent*.

1.3 Dimension

We say that a point is *smaller* than a line, a line is smaller than a plane, a plane smaller than Space. The Nothing is the smallest element, smaller than the point. This is expressed with the concept of 'dimension': to each element we assign a number, called its *dimension*, in the following way:

Element	Dimension
Space	3
plane	2
line	1
point	0
Nothing	−1

A point can be on a line, in which case we say that it is *contained* in the line, or that the line contains the point. A line can (completely) lie in a plane, in which case we say that the line is contained in the plane, or that the plane contains the line. All points, lines and planes are contained in space, the Nothing is contained in each element. For convenience we also admit that each element contains itself. So a point A contains point A, and also the Nothing. Line l contains all its points, but also l and the Nothing, etc.

If one element contains a second one, the dimension of the first must be bigger than or equal to that of the second. The converse is not true, however: if A is a point *not* on line l, l does not contain A, whereas its dimension is bigger than that of A.

1.4 Non-metrical qualities

Pure projective geometry is especially concerned with the non-metrical qualities of space. The foremost of these is that of *incidence*; a point which lies in a line is said to be incident with that line, etc. 'Incidence' is used for 'contains' as well as for 'is contained in.' Other non-metrical concepts are *join* and *meet*, see Section 1.6. We shall find that such qualities are much more deeply inherent in the nature of things than any considerations of size, or even ratios of sizes.

1.5 The infinite

Since we are not dealing with metrical things we shall not normally take notice of parts of lines or planes, which, being finite in length, immediately imply metrical considerations. We shall be interested in the whole, infinite entity. We must therefore broaden our ideas of what a line is really like. Let us consider two lines a and b lying in one plane (see Figure 1.3). Our imagination assures us that they will meet in one point, and one only. Let this point ab be called X.

Let O be any point of a, and let a turn with constant speed around O. It is clear that X will move along b with increasing speed, until a moment will come, when a and b are nearly parallel, when X is very far away, and rushing into the far distance at enormous speed. A short moment later, when the parallel position has been passed, X will be found at the 'other end' of b, and rushing inwards at like speed. Between these two moments is the instantaneous moment when the lines are parallel. Are we to tolerate that X exists, somewhere throughout the whole process, except for this instantaneous moment?

Modern geometry insists that we find a place for X even in this

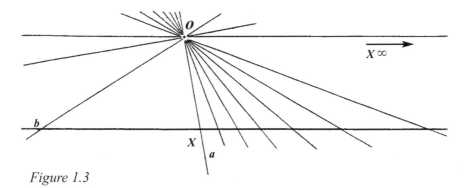

Figure 1.3

moment. We call it the *point at infinity* of the line, and we consider it to be simultaneously, one and the same point, to right and to left. The line has no 'other end'; it is a cyclic entity, — but still *straight* throughout its length.

We can, of course, imagine infinitudes of lines on the surface of a plane, and each will have its single point at infinity. On what locus do all these points at infinity lie? Consider two planes in space (Figure 1.4) that meet in a straight line x. Now suppose the grey plane turns around some line o which is parallel to the white plane. After a while it is at the position of the black plane (Figure 1.5) which meets the white bottom plane in another straight line, y. Going on, the meeting line will move to the right, with increasing speed, until the turning plane is in the position of the upper white plane,

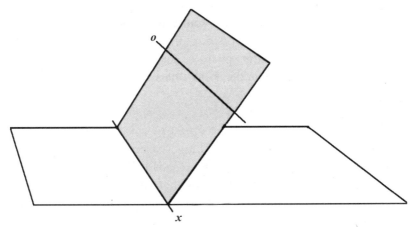

Figure 1.4

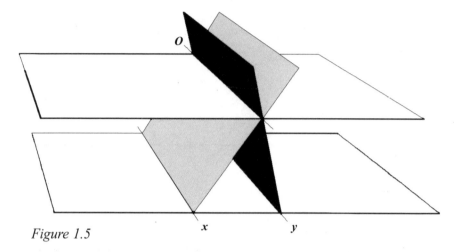

Figure 1.5

which is parallel to the lower one. Going a little bit further, the meeting line will reappear from the left. Again, are we to tolerate that the meeting line exists throughout the whole process, except for this parallel position? We accept that parallel planes share a (straight!) line, which is called the *line at infinity* of these planes. It is the infinite *horizon* of the planes, which *seems* to us like a vast circle. But it can't be a circle, because opposite points coincide.

Further, we can imagine infinitudes of lines and planes in space, and each will have its point or line at infinity. On what locus do all these points and lines at infinity lie? A moment of reflection teaches us that any two different lines at infinity meet in a point at infinity. Just because their generating planes meet in parallel lines (in Figure 1.6: *a, b, c, d*), these lines share one point at infinity (*P*).

So all the lines at infinity are contained in one flat plane. This is called the *plane at infinity* of Space. It is the infinitely distant plane which seems to us like the sphere of the starry heavens. But again, since opposite points coincide, it can't be a sphere.

1.6 Meet and join

The *meet* of two elements is the biggest element contained in both these elements. So, the meet of two planes is their common line (we have seen that in projective geometry there is always one). The meet of a plane and a line not in that plane is a point. The meet of two different lines is either a point (if they are contained in one plane); or it is the Nothing, in which case they are called *skew*.

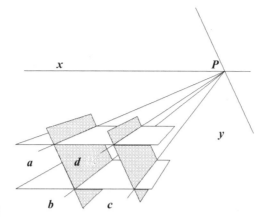

Figure 1.6

The *join* of two elements is the smallest element that contains both. So the join of two points is the line that joins them. The join of a line and a point not on that line is a plane. The join of two different lines is either a plane (if they meet in a point); or it is Space, in which case they are skew again.

In general the meet (or *intersection*) of two figures is the set of their common *elements*. So the meet of a sphere and a plane through its centre is a circle.

1.7 The dimension theorem

If we take two arbitrary elements and determine their meet and join, we discover a most interesting relationship of their dimensions.

element 1	element 2	meet	join
A	*l*	*A*	*l*
0	1	0	1

Consider a point *A* on a line *l*. Their meet is the point *A*, because *A* contains *A* (see Section 1.3) and *l* contains *A*, and there is no bigger element that is contained in both. Similarly their join is line *l*.

The last line of the table contains — of course — the dimension of the element in the middle line.

Next take a point *B* not on *l*. Now the meet is Nothing and the join is the plane α that contains them both.

element 1	element 2	meet	join
B	*l*	*Nothing*	α
0	1	-1	2

Take a line *m* and a plane not containing *m*. Their meet is the meeting point *C*, their join is Space.

element 1	element 2	meet	join
m	α	*C*	Space
1	2	0	3

Exercise 1a

Make such tables for all other possible configurations of two elements:
— point in plane
— point not in plane
— line in plane
Try to find the relationship between the dimensions. ❏

Exercise 1b

Make a table for each pair containing Nothing or Space. For instance:

element 1	element 2	meet	join
A	Nothing	Nothing	*A*
0	-1	-1	0

Does the relationship still hold?

If everything went well you discovered that adding the first two numbers always gives the same result as adding the last two. This is a very profound property of space; it is called the *dimension theorem* and it is by no means easy to prove it from the axioms:

> *The sum of the dimensions of any pair of elements equals the sum of the dimensions of their meet and join.*

2. Duality

2.1 Duality in space

Here are six facts which may be taken as being directly accessible to the imagination and not requiring proof. They are called the propositions of incidence.

The join of two different points is a line.	The meet of two different planes is a line.
The join of a line and a point which is not contained in that line, is a plane.	The meet of a line and a plane which does not contain that line, is a point.
If the meet of two lines is a point, their join is a plane; else it is Space.	If the join of two lines is a plane, their meet is a point; else it is Nothing.

When these facts are arranged in pairs like this, one is able to see a remarkable correspondence between them. One has to interchange
— the words *Space* and *Nothing*
— the words *point* and *plane*,
— the words *meet* and *join* and
— the expressions *contains* and *is contained in*
in one of the left-hand propositions, leaving the word 'line' where it is, and one arrives at the corresponding right-hand proposition.

This is known as the *principle of duality* and each proposition is called the *dual* of the other. The process of applying this transformation is known as *dualizing*.

2.2 Duality in the plane

If instead of working in space one confines oneself to the plane, the basic entities which one has to deal with are just the point and the line. The propositions of incidence are then these two:

The join of two points is a line.	The meet of two lines is a point.

To dualize within the plane one has to interchange
— the words *Plane* and *Nothing*
— the words *point* and *line*,
— the words *meet* and *join* and
— the expressions *contains* and *is contained in*.

2.3 Duality in the point

Instead of confining oneself to a plane, as in plane geometry, it would be quite feasible to confine oneself to a point and all the lines and points passing through it. Then one deals with what is called the *geometry of the point*. The basic entities with which one would deal would be the plane and the line. The propositions of incidence would then be these:

The meet of two planes is a line.	The join of two lines is a plane.

To dualize within the point one would interchange *plane* and *line* etc.

2.4 Polarity

If in some figure we dualize according to some prescribed rule, one speaks of *polarizing*, and the new figure is called the *polar* or *polar opposite* of the first.

As an example we will draw the polar of a cube. Draw first a perspective picture of a cube. It is most conveniently done by drawing all parallel edges to be actually parallel on the paper; that is, a cube as seen from an infinite distance (Figure 2.1). If now we find the centre of each face of the cube (by drawing its diagonals) we shall have a point for each plane of the cube. These points must be joined together in such a way that wherever two faces of the cube meet in a line of the cube, their corresponding points must be joined by a line. The resulting figure is an *octahedron*, the polar opposite of the cube.

Notice that the cube has: 6 planes, 12 lines, 8 points,
and the octahedron has: 6 points, 12 lines, 8 planes.

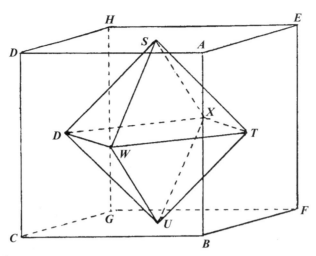

Figure 2.1

We have arranged that every plane of the cube has a point of the octahedron, and we now find that the qualities of space have ensured that every point of the cube has its corresponding plane of the octahedron; for instance, point *A* of the cube, where three lines meet, corresponds to plane *SWT* of the octahedron, in which three lines of the octahedron lie. Also we note that for every line of the cube there is a corresponding line of the octahedron. For instance, line *AB* is the meet of two planes of the cube. These planes have their corresponding points of the octahedron in *T* and *W*. Therefore *TW* is the line corresponding to *AB*, etc.

Notice plane *SWUX* which cuts from front to back through the centre of the octahedron; where is its corresponding point in the cube? Since the four lines *SW, WU, UX* and *XS* all lie in one plane, their corresponding lines must all meet in one point, which is the point we want.

Notice that the points *S* and *W* correspond to the top and front planes of the cube respectively; therefore the common line of *S* and *W* will correspond to the common line of these two planes; that is, the line *AD*. If similarly you now find the lines corresponding to the other three lines, *WU, UX* and *XS*, you will find that they are all parallel. But since the lines *SW, WU, UX* and *XS* are coplanar, our reasoning tells us that their corresponding lines should all share a common point. So we see that here, right at the start of our studies, we come on a situation where our reasoning can only make sense if we agree that parallel lines share a common point.

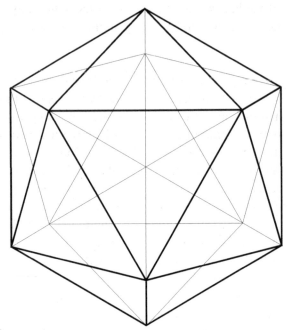

Figure 2.2

Now we find the corresponding points to the planes *SVUT* and *VWTX*, which pass through the middle of the octahedron. These three planes of the octahedron meet in its centre point. Now find the plane of the cube which corresponds to the central point of the octahedron. Remember that any three points which are not in line determine a plane; in this particular case the three points in question are all at infinity; so we see that our reasoning only makes sense if we agree that the infinite periphery has the form of a plane.

We find that if we regard the octahedron as being a solid body, made let us say of atoms (points), its polar form is a hollow cube, composed of planes. Every point on the surface of, let us say, the plane *SWT* of the octahedron will have its answering plane passing through point *A* of the cube. Every point under the surface of the octahedron will have its corresponding plane in space outside the cube, and the centre point of the octahedron will correspond to the plane at infinity. The planes of the cube will fill all space except the hollow part within it, and its central plane will be the plane at infinity. We see that when we polarize we change not only the outward shape, but also the very quality of the figure; even our ideas of inside and outside have to undergo revision.

Of course if we had regarded the cube as the solid body we should have found the octahedron to be hollow in the same way.

Exercise 2a

Draw a perspective picture of an *icosahedron* and its polar, using the same principle as before: replace each plane by the centre of its triangle.

The icosahedron can be constructed in the following way.

Draw a circle of 7 cm radius and inscribe a regular hexagon. Draw a circle with the same centre and radius 5 cm. Join opposite points of the hexagon and where these lines meet the smaller circle alternately, draw an isosceles triangle, as in Figure 2.2. Now join each point of this triangle to the three nearest points of the hexagon. These lines show all the front of the icosahedron. By drawing another set exactly the same, using the triangle which is the other way up, one draws all the lines of the back of the body (thin in the figure). Find the midpoint of each face (it is the meet of the medians) and when these are joined up according to the rule of the example of the cube, the polar figure (a *dodecahedron*) will appear.

Work carefully through this figure and the previous one, identifying all corresponding lines, points and planes.

2.5 Polarizing a plane curve

Now let us try polarizing within the plane — point to line and line to point. We will take the case of a simple curve, see Figure 2.3. Of this we may say, that of all the infinitudes of points in the plane we have chosen a single infinitude; that is, those points which obey some given relation or law. (What this relation or law is for this curve will be discussed in Section 6.8.) This infinitude of points, lying side by side, constitutes all that we can ordinarily see of the curve (we will see in Chapter 14 that it in fact belongs to a much richer organism of movement). We will call it a *locus* or *pointwise curve*.

Polarizing, we say that of all the infinitudes of lines lying within the plane we have chosen a single infinitude which obeys the polar relationship. Since the law that defines this curve will not be given until later, we are not in a position to rigorously polarize this figure. But just from looking at the figure and applying the principle of duality, we can get a pretty good sense of what the polar must look like.

Given any two points of the pointwise curve, they lie on a line

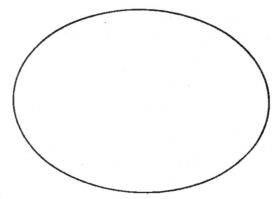

Figure 2.3

(this must be so by one of the propositions of incidence). Note that we have chosen our curve so that no other point of it lies on that line.

The dual statement to this is: given any two lines of the *linewise curve*, they meet in a point (this must be so by the dual proposition of incidence). No other line of the linewise curve must go through this point.

In addition, it suffices to notice that our pointwise curve is cyclical: a point moving along it comes back to its initial position after having traversed all the points of the curve exactly once.

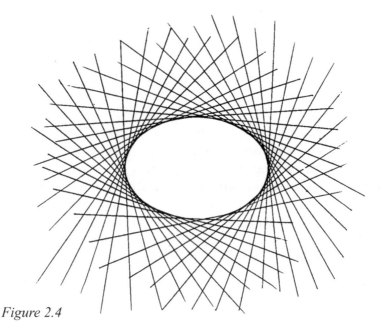

Figure 2.4

Thus, dually, a line moving 'along' the linewise curve must come back to its initial position after having traversed all the lines of the linewise curve exactly once.

The reader is encouraged to ponder these stated properties that the polar curve should have, and try to come up with a linewise curve satisfying the two, before reading on. There is no one answer. In fact there is a large class of linewise curves that satisfy the two properties. (This is true even if we knew the precise law of our pointwise curve, and thus can polarize it as well. However then the class would be considerably smaller. That is done in Chapter 6 and 7.)

Figure 2.4 shows a polar of our pointwise curve. We have simply drawn the tangent envelope to our original curve. However there is as yet no reason to pick one shape over another, if both shapes satisfy our two properties.

2.6 Polarizing a plane curve in space

We could have proceeded differently and polarized our original pointwise curve (Figure 2.3) in *space*; that is, point to plane and plane to point, with the line as the intermediate element. Since Figure 2.3 is a curve of points lying in a plane, the polar would be a 'curve' of planes lying in a point. Applying the same kind of analysis we used for the planar case (the reader is encouraged to do this), we are led to something similar to a cone formed of its tangent planes.

This is difficult to illustrate. Depicted in Figure 2.5 is a cone with one tangent plane. Imagine this plane turning in the apex point of the cone, so that it is always tangent to the cone, and now imagine the whole cone generated by such a movement of a plane. You then have the figure required.

Polarizing Figure 2.4 similarly, we arrive at something similar to a cone of lines lying in a point; see Figure 2.6. (In these, and all figures one must imagine all lines and planes extending to infinity. Sometimes they have to be cut off at finite distances in order to make the forms visible to the eye.)

When we decide to study plane geometry we confine ourselves to one particular plane, and we then have only points and lines, as it were, to play with. But it would be equally feasible to work out a point geometry in which we confine ourselves to the planes and lines passing through some particular point.

The ordering of planes and lines in our figures would then best be

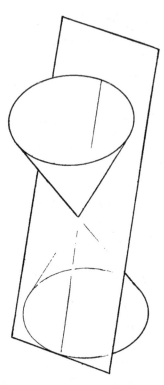

Figure 2.5

regarded not as three dimensional figures in space, but as two
dimensional figures experienced intensively in the way they pass
through the point: a line passing through the point would be thought of
as the quintessence of infinitesimal straightness in a particular direction; a plane as the quintessence of infinitesimal flatness in a particular
position. (It is possible to make such a consideration formal with the aid
of ideas from calculus, but we will not do this.) In thinking of it this
way, we find that indeed the whole of plane geometry is literally contained within just one point. Now the space polar to this geometry is just
the geometry of points and lines in a plane, which is a *two*-dimensional
geometry (two dimensions means two independent degrees of freedom). But since dualization is a process that does not add or subtract
anything — in the case of space-dualization just interchanging the role
of points and planes — our microcosmic geometry that is totally contained within a point must also be a two-dimensional geometry! A complete geometry, in every proposition dual to the ordinary plane
geometry, could thus be worked out in this way. For example, as previously stated, to dualize *within* the point one would simply interchange
plane and line.

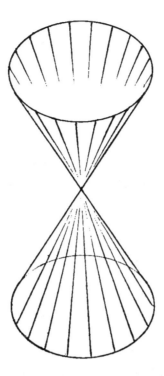

Figure 2.6

Since the space dualization process preserves incidence, the incidence geometry of points and lines in a plane is *logically* totally equivalent to the incidence geometry of planes and lines in a point, except that the words 'point' and 'plane' must everywhere be interchanged. But to our imagination the two are utterly different! As we proceed in this book with plane geometry it can frequently not only be interesting but also insightful to space-dualize any statement to obtain the space-dual in point geometry. Since man is a being not only logical but also of imagination, the difference in quality between the extensive plane geometry and the intensive point geometry can be of great importance.

One particularly nice feature of point geometry is that clearly all lines and all planes in it are of equal stature. Not one is singled out, as appears to be the case with the line and points at infinity in plane geometry. (It only *seems* so, since — as we mentioned in the introduction — pure projective geometry is concerned with non-metrical properties of space, whereas the concept of infinity is metrical. A very good way to see that the line at infinity is, in fact, no more special in pure projective geometry than any other line is to see what happens in the logically equivalent point geometry!)

Leaving point geometry aside for now and regarding things extensively, we could have arrived at Figure 2.5 in a different way. Let us
imagine Figure 2.3, the pointwise curve in a plane, and another point
somewhere in space outside the plane. By joining every point of the
curve to this point we would arrive at the linewise cone. Therefore we
say that Figure 2.6 arises by projection from Figure 2.3. Similarly we
can get Figure 2.5 from Figure 2.4.

We could tabulate our four figures as shown

Pointwise Curve (Fig. 2.3) ----------------------- Linewise Curve (Fig. 2.4)

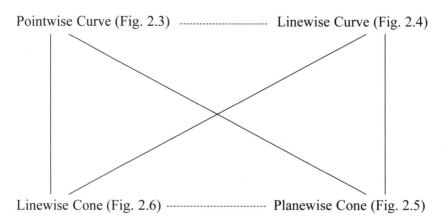

Linewise Cone (Fig. 2.6) ----------------------- Planewise Cone (Fig. 2.5)

This table is called a *commutative diagram*, because the route one
takes in going from one place in the figure to another does not affect
the outcome in the sense that what becomes points by one route will
also become points by another route, and the two sets of points have
the same properties of incidence. Similarly for lines and planes.

This table can be very helpful. For example, going from plane
geometry (Figure 2.3) to point geometry (Figure 2.5) is very easily
achieved when one has become proficient in polarizing within the
plane: simply go from Figure 2.3 to Figure 2.5 via Figure 2.4. (Of
course, if one is more proficient in polarizing in space, one should go
directly from Figure 2.3 to Figure 2.5.)

It was mentioned before that it can be interesting and instructive to
translate all statements in plane geometry to statements in point geometry, and this can be achieved by the two above-mentioned ways as
well as by going from Figure 2.3 to Figure 2.5 via Figure 2.6.
However, many of the insights that can be gained from this can also be
gained by simply going from Figure 2.3 to Figure 2.6. In particular, the
line and points at infinity of Figure 2.3 no longer appear special when
translated into Figure 2.6.

If one wants to then go from Figure 2.6 to Figure 2.5, this is particularly easy if one chooses the polar of a line through our point to be its perpendicular plane through that point and vice versa.

2.7 An example of dualizing in the plane

It takes some practice before one can dualize fluently. In the early stages, or at any time when difficulty is experienced, it is a help to write out the stages of one's original construction or proposition naming every point and line. Then write out the dual to each stage and make the construction. Here is an example.

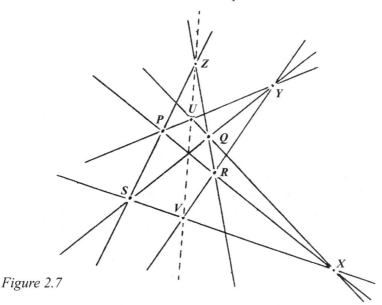

Figure 2.7

Original proposition

Let PQRS be a quadrangle and X and Y be any points on PR and QS respectively. QX and PY meet in U and SX and RY meet in V. It will be found that the line UV passes through the meet of PS and QR at Z (Figure 2.7).

1) Mark four points *P Q R S* of a quadrangle.
2) Let the common point of line *PS* and line *QR* be point Z.
3) Mark point *X* on line *PR* and point *Y* on line *QS*.
4) Draw lines *XQ* and *YP* and let them meet in point *U*.
5) Draw lines *XS* and *YR* and let them meet in point *V*.
6) Points *U V Z* are collinear.

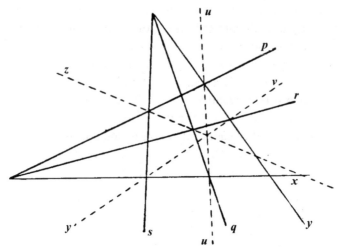

Figure 2.8

The dual (Figure 2.8).
  1) Draw four lines *p q r s* of a quadrilateral.
  2) Let the common line of points *ps* and *qr* be *z*.
  3) Draw line *x* through point *pr* and line *y* through point *qs*.
  4) Mark points *xq* and *yp* and let their joining line be *u*.
  5) Mark points *xs* and *yr* and let their joining line be *v*.
  6) Lines *u v z* are concurrent.

Exercise 2b

On two lines *x* and *x'* mark three points each, *A B C* and *A' B' C'*. Let *AB'* and *A'B* meet in *M*; *AC'* and *A'C* in *N*; *BC'* and *B'C* in *L*. It will be found that *L, M* and *N* are in line. Dualize. This is the well-known *Theorem of Pappos*.*

---

* The Theorem of Pappos is a very profound one. In the axiomatic building up of projective geometry, however strange it may seem, the proposition of Pappos is equivalent to the property $ab = ba$ for numbers (for instance, 4 times 3 = 3 times 4).

Exercise 2c

Let *ABC* be any triangle and *O* be any point (conveniently though not necessarily, within the triangle). *AO, BO* and *CO* meet the opposite sides of the triangle in *A'*, *B'*, *C'* respectively. A line through *A* meets *A'B'* and *A'C'* in *B"* and *C"* respectively. It will be found that *BB"*, *CC"* and *B'C'* are concurrent. Dualize.

We shall find many other cases to dualize as the course proceeds.

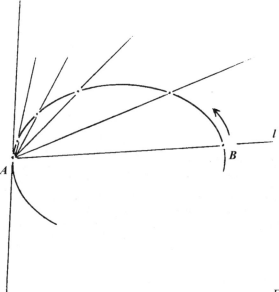

Figure 2.9

2.8 Tangency

The projective definition of a tangent to a curve depends on qualities of incidence. Let a line *l* meet a curve in two points *A* and *B* (Figure 2.9). Now let the line turn around point *A* in such a way that point *B* moves towards *A*. A moment will obviously come when *B* coincides with *A*. At this moment we say that the line *l* is *tangent* to the curve. A tangent to a curve is therefore any line which meets it in two coincident points.

Exercise 2d

Dualize the above definition and Figure 2.9. We then arrive at a definition for the point at which a tangent touches its linewise curve.

2.9 Order and class

Curves are arranged in *orders* and *classes*, according to their form. The *order* of a curve tells us how many points this curve has in common with any line of its plane. This relationship must be seen as existing between the curve and every line whatever in its plane. For instance, a circle is a curve of second order; and we can easily find many lines which can be seen to meet it in two points; but we must also understand that a line which passes 'outside' such a curve, and apparently does not meet it, also shares two points with the curve. These points, which are by their nature invisible, are called *imaginary points*. At this juncture we can only mention the existence of these points, but later in this book we shall see quite easily how such points must be envisaged.

Similarly the *class* of a curve is the number of tangent lines, imaginary or otherwise, which each point of the plane sends to it.

Exercise 2e

By dualizing in space, find definitions for cones of second order and of second class. To do the latter, we need to think of the second order curve thus: we have a curve of points in a plane π, which is such that there are lines of plane π which share two points in common with the curve.

Now the dual statement will begin. We have a cone of planes in a point P, which is such that ... etc. And when this statement is finished it will, of course, refer to the class of the cone, since we started with the order of the curve. Order and class are dual to one another. Notice that if we have a second class cone, and we cut it by any other plane, the resulting section will be a second class curve. But if we dualize it in space (point to plane, plane to point, line to line) we arrive at a second order curve. If, however, we dualize it within the point (line to plane and plane to line) then we get a second order cone.

Figure 2.10

2.10 Special features

Apart from just 'going round' there are a number of special things a curve can do, and of these there are four chief types. The first two are a *cusp* and an *inflexion* (or a *flex*), see Figure 2.10.

These are dual to one another. One can see that whereas a point moving round the curve would have to come momentarily to a stop in passing through a cusp and then would have to reverse the direction of its movement, a tangent to the curve would slide easily and smoothly through. On the other hand, a point slides easily through an inflexion but a tangent which is turning in one direction has to stop its turning and then start turning in the other direction.

More precisely, consider the tangent to the curve at the cusp and any point P on this tangent (Figure 2.11).

We see that this point has three tangents to the curve — that to the

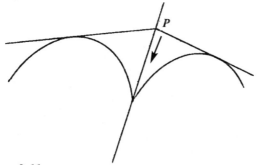

Figure 2.11

cusp and two others. Now let the point move along the tangent towards the cusp. We see that the other two tangents gradually get closer to the cusp tangent and in the moment that the point reaches the curve, the three tangents will coincide. Thus we may say that the point which carries three coincident tangents to the curve lies on the cusp.

Exercise 2f
Dualize Figure 2.11 and find a definition for the tangent at an inflexion.

Figure 2.12

The other two special features are the *node* and the *bitangent* (Figure 2.12). Clearly these are dual to one another, the node being a point through which the pointwise curve passes twice, and the bitangent being a line through which the linewise curve passes twice.

3. Collineation

3.1 Desargues' triangle theorem

Let us suppose that we have two triangles ABC and $A'B'C'$ situated more or less like those in Figure 3.1.

We notice that the common lines of corresponding points AA', BB' and CC' form a little triangle, while the common points of pairs of corresponding sides aa', bb' and cc' form a long thin triangle. By changing the two triangles very slightly, one could arrange that the little triangle of three lines AA', BB' and CC' shrinks into a point; one would then find that the long thin triangle aa' bb' cc' melts into one line (Figure 3.2).

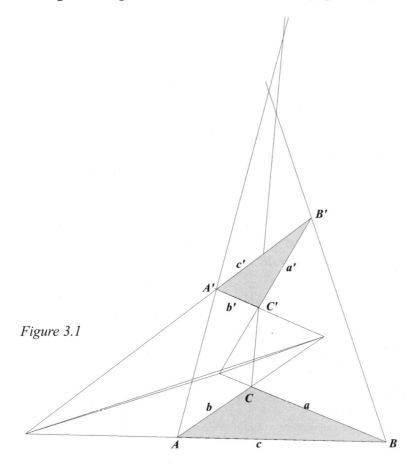

Figure 3.1

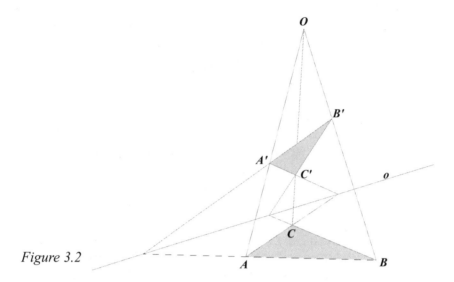

Figure 3.2

This is *Desargues' Triangle Theorem*, which may be stated thus:

> *If two triangles are such that the lines common to*
> *corresponding points meet in a point, then the points*
> *common to corresponding lines lie on a line.*

It is interesting to note that this theorem cannot be proved from the propositions of incidence within the plane. However, as soon as we see the theorem in the context of three dimensions its truth becomes readily accessible to the imagination.*

Consider Figure 3.2. Imagine a trihedron with its apex at point *O*. This will be lines *AA'*, *BB'* and *CC'*, imagining, let us say, *CC'* the centre one to be little behind the other two. This trihedron is cut by a plane and its section is the triangle *ABC*.

Now the trihedron is also cut by another plane and on that plane its section is the triangle *A'B'C'*. These two triangles are now, as pictured in perspective, in the Desargues' relation — that is, the common lines of corresponding pairs of points all meet at a point, *O*.

Now consider the lines *AB* and *A'B'*. We must first assure ourselves that they do in fact meet, for now that we are working in space, two

---

* It is possible to 'invent' a 'projective' plane geometry in which the Desargues proposition does not hold. It is a weird, artificial geometry. But its existence emphasizes the importance and profoundness of the theorem. Also it suggests that plane projective geometry can hardly exist without a third dimension.

lines do not necessarily meet — they may be skew. However we see that these two lines have a plane in common, the front plane of the trihedron. Therefore, by the propositions of incidence they must also have a point in common. Similarly we can assure ourselves that BC and $B'C'$, as well as CA and $C'A'$, also meet; they do not pass one behind the other only appearing to meet in our figure.

But, again by the propositions of incidence, any two distinct planes meet in a line. Therefore the meeting points of corresponding sides must all lie on a line, o, the common line of the two planes which contain those sides.

We see that Desargues' theorem is in fact just a two dimensional picture of a three dimensional truth which is almost self-evident.*

3.2 Shadow drawing

From this we can get a useful method of shadow drawing in perspective. We can imagine that point O is a source of light. Then the triangle ABC becomes the shadow of triangle $A'B'C'$. We see that every line connecting a point with its shadow passes through one point (O, the source of light) and that each line and its shadow, produced if necessary, will meet in a point which lies on one particular line (the common line, o, of the two planes).

As an example, draw a horizontal line l to represent the meet of the horizontal plane (the ground) with a vertical plane. High up in the figure place a point O, to represent the sun and in the vertical plane draw any object you wish.

Now select some point of the object, A, and draw the line OA. As soon as you fixed point O you sacrificed a part of your freedom, for the shadow A' of A must lie somewhere along the line OA. You are still free to decide where, so now place A' in a convenient position. It is helpful in drawing this figure:

1   to have the object, whose shadow you are seeking, slightly above line l,
2   to make the object a simple two-dimensional form — say a cross, or five-pointed star, or some other simple form
3   to place point A at the very top of the object, and

---

* To complete the proof one has to show that there exists a space configuration which is perspective with the plane one, or equivalently, one has to find a triangle outside the plane which is perspective with both ones in the plane. This, however, is straightforward.

4   to place its shadow, A', on the other side of line l, low down on
     the page along the line OA.

Having done so you have no more freedom left; the whole phenomenon, in relation to your position as viewer, is now fixed.

Next take any other point of the object, B. Draw OB, and, of course,
B' must lie somewhere along this line. Now draw line AB, it meets line
l in P. B' must also be on PA'. So the meeting point of PA' and OB is
B', which is the shadow of B. In the same way all the rest of the shadow
may be constructed.

Notice that having found B' in this way, you may proceed to find the
point C' corresponding to some C, by the same process. But you can also
find point C' by using points B and B' in the same way that you used A
and A' and every time the same point C' will result. One finds here a wonderful spatial anastamosis. In doing a large figure one often needs, for the
sake of convenience, to change the pair of points which one is using.

Exercise 3a

Repeat the above construction using a circle for the object in the vertical plane. Obviously the light streaming through this circle forms a
cone and this is caught on the second plane as a conic section.

3.3 Collineation

The construction of the previous section is something more far-
reaching than just a method of drawing shadows. Let us forget the
three dimensional aspect for a moment and look at it as just a plane
figure. We find that we have here a *projective transformation* of the
plane within itself. We may say that we transform the plane into
itself, so that A transforms into A', B into B', etc. We see that the
transformation is one-to-one, that is to say, each point transforms
uniquely into one point only. *Projective* means that incidence is kept:
if some point X is on a line y before the transformation, after the
transformation the image point X' must be on the image line y'.

A projective transformation which turns points into points, lines into
lines and, in space, planes into planes, one to one, is in general known
as a *collineation*. Clearly the order and class of any curve remains
unchanged in any collineation.*

---

* The other type of projective transformation is *correlation* or *polarity,* which
    turns points into planes and vice versa, and lines into lines; or, in the plane,
    which turns points into lines and vice versa.

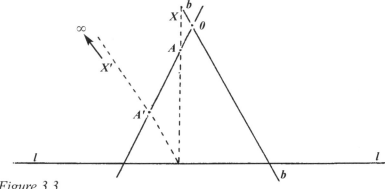

Figure 3.3

Notice that there is a line of self-corresponding points, line *l*, called the *axis* of the collineation, and a point of *self-corresponding lines*, point *O*, its *centre*. But the points of these self-corresponding lines are not themselves self-corresponding; each transforms into another point of the same line, excepting of course, point *O* and the common point with line *l*. Similarly the lines of the self-corresponding points of line *l* are not themselves self-corresponding, but each transforms into another line of the same point, excepting of course, line *l* and the common line with point *O*. It can be seen that the functions of point *O* and of line *l* are in every way dual to one another. The transformation with which we are dealing here is known as *homology*, and it is a special case of a collineation, viz. a collineation with a line of self-corresponding points and a point of self-corresponding lines.

Let us establish a homology, as in Figure 3.3, the self-corresponding elements being point *O* and line *l*.

Note that it is not necessary for *A'* to be on the other side of line *l* from *A*. Consider any other line *b*, through point *O*. We wish to find the point *X* of this line which will transform into the point at infinity *X'* of the line. We draw the line *X'A'* parallel to *b*, and from where this meets line *l* we draw a line through point *A* to meet *b* in *X*. Now for all the other lines through *O* we can find their points *X* which will transform into their points at infinity. If the infinite points of a plane really lie on a straight line, these points *X* ought to be found to be collinear. We can say that their line (let us call it *x*) transforms into the line at infinity. But a line and that into which it transforms are bound to meet on line *l*. In other words, lines *x* and *l* should meet at infinity, i e. they should be parallel.

Exercise 3b

Carry out the above construction using several lines through O and confirm that line x appears in the way in which we have reasoned that it should. There will also be a line into which the line at infinity transforms. What can we say immediately about this line? Find a construction to draw it.

In traditional geometry the family of second order curves, usually called *conics*, is divided into three main types of curve, called *ellipse*, *hyperbola* and *parabola*. The last of these is specially familiar to our eyes since it is the path followed by any object when it is thrown. In projective geometry we come to the rather surprising definition of a parabola that it is any conic which is tangent to the line at infinity. It is not possible at this stage to go into the reasons for this; let us just take it as a given fact, and use it to work the next exercise.

Exercise 3c

Construct a parabola. We know that a circle subjected to homology transforms into a conic. The preceding work has shown us how to find that line of the plane which will transform into the line at infinity. The property of tangency is a projective one — that is, it relies on considerations of incidence — and is thus preserved by projective transformation. Therefore we can reason that if we put a circle tangent to this line, it will transform into a parabola. It is a useful exercise to try it, using circles of different sizes, both above and below the tangent line.

It is not necessary for O to be away from line l; it may lie on it and the whole process of transformation continues exactly as before. In such a case the homology is known as an *elation*. It is good practice to make an elation, carrying out the construction in just the same way as for the ordinary homology.

Exercise 3d

Do this.

Two cases, one of homology, and one of elation, are specially of interest.
    In the first case (Figure 3.4) line l is at a finite position and O is at infinity in a direction at right angles to l. We fix A and A' so that the line AA' passes through O. We can then construct B' corresponding to any point B,

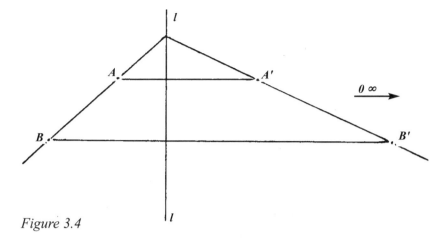

Figure 3.4

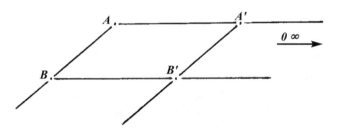

Figure 3.5

as in the figure. We see that the transformed picture is *similar* to the original one, but 'stretched' away from line *l* by a constant ratio.

In the second case (Figure 3.5), we have an elation in which line *l* is the line at infinity and, of course, *O* is also at infinity.

We see that the whole plane is moved bodily, in the same direction and the same distance as the original pair, *A* and *A'*. In other words, whenever we move any rigid body in a straight line from one position to another, every point of it is obeying the laws of a parallel elation, all the self-corresponding elements of which are at infinity. We see a first indication that this infinitely distant line is indeed as absolute, from which our physical world takes the laws of its being. This elation is called a *translation*.

Exercise 3e

Investigate a homology in which the centre is finite and the axis is at infinity. Such a homology is called a *multiplication*. Why?

4. Cross Ratio

4.1 Cross ratio

The simplest case of projection which we find is when a range of points is projected by the rays of some point from one line to another (Figure 4.1).

When the range $ABC...$ of line x is projected from O into range $A'B'C'...$ of line x', projective geometry is concerned to know which properties of the range remain invariant. Obviously such properties must be more deeply inherent in the nature of things than those which disappear or change during projection.

A few moments' experiment with a figure will assure us that all considerations of size, and ratios of sizes, are changed. There is, however, one type of metrical quality which will be found to remain unchanged in any projection or series of projections.

Consider the ratios $AB:BC$ and $A'B':B'C'$. Each of these can be expressed as a pure number, quite regardless of the units used in measurement, and (unless we have done something special to ensure that they remain unchanged, like making x and x' parallel) they will surely be different from one another. Similarly, the ratios $AD:DC$ and $A'D':D'C'$ will usually be different. Consider however, the double ratio which we may write like this

$$\frac{AB}{BC} : \frac{AD}{DC}$$

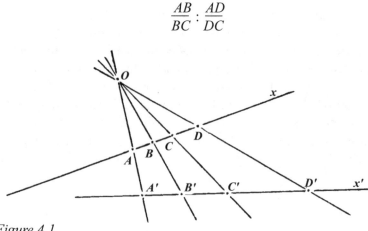

Figure 4.1

It is known as a *cross ratio* and it appears to remain the same after projection.

Exercise 4a

Measure and calculate this ratio for the points *ABCD* as well as for *A'B'C'D'*. Verify that they are (practically) equal.

There are 24 different ways of grouping the mutual distances between any four given points to give a cross ratio. The different values of the cross ratios of any given four points, however, are only six in number, there being four of the possible groups to each value. All the different ways of finding a cross ratio are projectively equivalent and there is no use, for our purposes, in remembering more than one. If we decide on one, and stick to it in all our working, then we can prove two things:

1) that any group of four points on a line, duly named in order, has one unique number associated with it (its cross ratio), although, of course, this number will also be associated with many other groups along the line.

2) that if we regard three points of a line as fixed, say *A*, *B* and *C*, then any fourth point, *D*, of the line, will be associated with a unique number; that is, the number of the cross-ratio of the four points, and that this number will be associated in this way *with no other point of the line*. In other words, there will be a one-to-one relationship between the numbers of our number system (extended by one infinite 'number,' ∞) and the points of the line.

Exercise 4b

Verify this. Calculate the cross ratio in several positions of *D*. If *D* is at infinity the cross ratio reduces to *AB/BC*. If *D* coincides with *A, B* or *C* the cross ratio will be ∞, 1 or 0 respectively.

Unless otherwise stated, by the cross ratio of the four points *A*, *B*, *C* and *D* in that order — denoted by (*ABCD*) — we will mean exclusively the cross ratio as defined above; so

$$(ABCD) = \frac{AB}{BC} : \frac{AD}{DC}$$

This is easy to remember; in the first fraction we take a journey from A to C calling at B, and in the second we go from A to C calling at D.*

In making any measures along a line one can normally choose which direction one will consider to be positive and then, of course, the other will be negative. In the case of our cross ratio we see that if we consider AB to be positive, then all the other measurements will be positive also, except the last, DC. This means that our cross ratios will normally be negative numbers, as long as the points are named in order, ABCD.

Even if one or more of the points pass through infinity, and 'come back from the other side,' giving an ordering like DABC, etc. it is easy to see that, A to B being considered positive, there will always be either one or three negative measures in our expression, making the final answer negative.

It will be found to be a fundamental quality of all projective transformations that they leave all cross ratios invariant. The fact that this subtle double ratio is so deeply inherent in the qualities of the line that it is never changed by any number of projective transformations is an exceedingly important one. We should never cease to wonder at it. We will not at the moment try to go into all the deep-reaching logical connections concerned with it, but it would be very worth while that the reader should draw one or two careful figures, and by exact measurement and calculation confirm this invariance under projection for himself.

It can also be shown that any transformation which has the quality of preserving all cross ratios, universally among all the elements of its sets, is a one-to-one projective correspondence. So:

A one-to-one correspondence is projective if, and only, if it leaves the cross ratio invariant.

Now let us consider four lines of a pencil, abcd (Figure 4.2). These will be cut by any line x, in four points ABCD, which will have a definite cross ratio. By what has gone before, it is apparent that no matter where line x may move to, the cross ratio of the four points in which abcd cut it, will remain constant.

We may say that this cross ratio is inherent in those four lines in such a way that wherever they may be cut by another line, that cross ratio immediately appears. We are justified therefore in speaking of

---

* Usually the cross ratio of ABCD is defined as $(AC/BC)/(AD/BD)$ which equals $1-x$ if x is the value according to the definition of Edwards. *(LdeB)*.

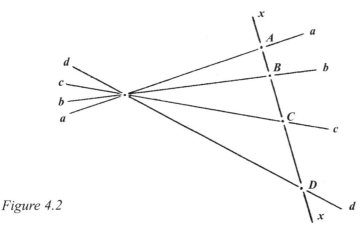

Figure 4.2

this cross ratio as belonging to that pencil of four lines, just as a cross ratio is said to belong to a range of four points.

For the reader familiar with elementary trigonometry, the proof of the invariance of the cross ratio should not be too difficult. Let h be the distance of the centre O of the pencil to our line x. Then

$$(ABCD) \;=\; \frac{AB}{BC} \cdot \frac{AD}{DC} = \frac{AB.\,DC}{AD.\,BC} = \frac{hAB/2 \,.\, hDC/2}{hAD/2 \,.\, hBC/2} = \frac{OAB.ODC}{OAD.OBC}$$

$$\frac{\tfrac{1}{2}OA.OBsin\angle AOB \;.\; \tfrac{1}{2}OD.OCsin\angle DOC}{\tfrac{1}{2}OA.ODsin\angle AOD \;.\; \tfrac{1}{2}OB.OCsin\angle BOC} = \frac{sin\angle AOB \;.\; sin\angle DOC}{sin\angle AOD \;.\; sin\angle BOC}$$

OAB meaning the area of the triangle. The result is independent of the position of x.

4.2 Terminology

When two lines are in a situation like x and x', in Figure 4.1, they are said to be in perspective one with another, and the transformation which carries A into A', B into B', etc. is called a *perspectivity*; x and x' are said to carry *perspective ranges* of points. Likewise, if we dualize Figure 4.1 we shall come to *perspective pencils* of lines, or we may say that there is a perspectivity between these pencils. We shall find that the invariance of cross ratio is a property common to all perspectivities.

If now we continue Figure 4.1 by carrying out a further perspectivity on line x' from a new raying point, on to a new line, x'', anywhere in the plane, we call the relation between the first line, x, and the last line, x'', a *projectivity*; and these lines are said to carry *projective ranges*. This relationship, a projectivity between two lines, is one of the very fundamental ones in our geometry. It can be established in various ways, but it can be shown that, however it was established, it can be broken down into a series of at most two perspectivities (see next exercise). Obviously, a projectivity shares with the perspectivity the quality of preserving all cross ratios unchanged, but in other respects its qualities are different, and we shall go on to study these in greater detail in Chapter 5.

Exercise 4c

Draw a line a with at least four points A, B, C and D on it (Figure 4.3). From some point O not on a, project these points on a second line a' (not through O) to get the points A', B', C' and D'. From a point P project the new points on to a third line a'' to get the points A'', B'', etc. Finally project these from a point Q on a fourth line a''' to get A''', B''' etc. (You could go on like this for a while.)

Now erase everything but the first line a and the last one a''' with their points. We will find one intermediate line b and two points R and S with which we can project a onto a''', A on A''' etc.

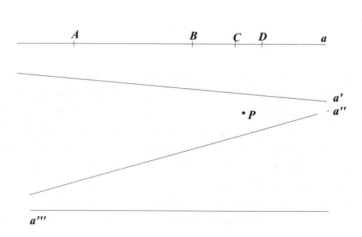

Figure 4.3

To achieve this take any point *R* on *AA'''* and any line *b* through *A'''*. Project *a* from *R* on *b* to get the points *A'''*, *E, F* and *G*. Let *S* be the meet of *B'''E* and *C'''F*. Project *b* from *S* onto *a'''* and verify that indeed *G* is projected on *D'''*. See also Section 6.3.

4.3 The four-point and the harmonic ratio

Now let us consider another figure. We take a line *x* and on it we mark three points, *A, C* and *D* (Figure 4.4).

The complete figure is given here for the sake of reference but you are strongly recommended to make your own, stage by stage, as you reads this.

Through *A* draw any two lines *p* and *q* and through *D* one line *m*. Let *m* meet *p* and *q* in *G* and *E*. Draw lines *CG* and *CE* and call them *r* and *s*. Let *r* and *q* meet in *H* and *s* and *p* in *F*. We have now constructed a quadrangle *EFGH*, which we call a *four-point* and a quadrilateral *pqrs* which we call a *four-line*. It must be stressed that this four-point and four-line, constructed in this way, is in no way different from any general four-point or four-line of any shape whatever; all the qualities we deduce about these figures belong to any four-point or four-line. The reasons for constructing them in this way are two — firstly, it ensures that the relevant points of our figure keep on the page and secondly, this method of construction will be seen to have useful applications later on.

Now consider carefully what we have — a four-point *EFGH*, and

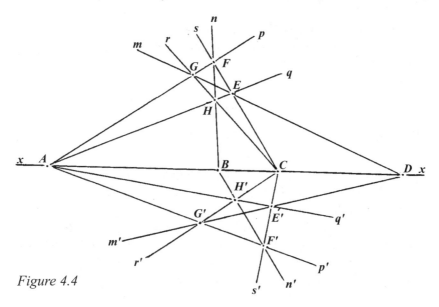

Figure 4.4

one diagonal *m*. Obviously, to complete the figure we need to draw the second diagonal *n*. Let it meet line *x* in *B*. Now we have four points along *x*, *ABCD*, and the four-point has paired them off. *A* and *C* are the meeting points of opposite *pairs* of sides of the four-point, while *B* and *D* are meeting points of the *single* diagonals.

Now repeat the construction below line *x*; starting as before from *A*, *C*, and *D*. We find, that no matter at what angle we make the lines *p'*, *q'*, *m'*, etc., and no matter what the shape of the four-point *E'F'G'H'* is, the second diagonal is bound to meet *x* at the same point *B*.

It is easy to see that this must be so. Consider triangles *EGH* and *E'G'H'*. We see that their corresponding lines meet on a line (at *A*, *C*, and *D*). Thus, they are in homology and lines *EE'*, *GG'* and *HH'* are concurrent. Similarly, we find that triangles *EFG* and *E'F'G'* are in homology, so that *EE'*, *FF'* and *GG'* are concurrent. This means that the two four-points are in homology and the corresponding lines *FH* and *F'H'* must meet on line *x*. Try to see as vividly as possible that the one figure is the shadow of the other.

We see that, given three points along line *x*, any four-point whatever, constructed on them in this way, will pick out one unique point *B* to pair with *D*. This should lead us to ask whether there is not, in fact, some special quality about four points which bear such a relation to a four-point.

We find in fact that, whatever positions and spacing we have chosen for the original points *A, C* and *D*, the cross ratio *ABCD* will be the same.

Exercise 4d

Draw a figure with these four points *ABCD* in this relationship, taking the spacing of the first three points quite arbitrarily. Then measure and calculate the cross ratio

$$\frac{AB}{BC} : \frac{AD}{DC}$$

You will find that it always comes to −1.

The fact that this must be so can be seen quite clearly from our figure. Let *O* be the meet of *EG* and *FH*. If we take *H* as a centre of projection, we see that the four points *ABCD* project into points *EOGD* respectively. Therefore the cross ratio of *EOGD* is equal to that of *ABCD*. But now projecting *EOGD* back on to line *x* from point *F* we find that these points project into points *CBAD*. Therefore the cross ratio of *CBAD* equals the cross ratio of *ABCD*. In other words:

$$\frac{AB}{BC} : \frac{AD}{DC} = \frac{CB}{BA} : \frac{DC}{DA}$$

Now we know that a ratio can be expressed as the first number divided by the second, so we might write:

$$\frac{AB}{BC} \times \frac{DC}{AD} = \frac{CB}{BA} \times \frac{DA}{CD}$$

In other words, the cross ratio *ABCD* equals its own reciprocal. This is only true of the numbers 1 and –1; we have already seen that this cross ratio is negative, so it must be –1.

This cross ratio is known as the *harmonic ratio* and the four points are said to be *harmonic*. Any harmonic ratio divides its four points into pairs. We say that the points *A* and *C* are harmonic *conjugates*, or mates, with respect to *B* and *D*, and similarly, that *B* and *D* are harmonic with respect to *A* and *C*.

Up to now, we have spoken about the diagonals of our four-point. This is not really a projective way of speaking. The points of the four-point are joined by six lines and no two of these are more truly diagonals than any other two. For instance, in Figure 4.5, which lines are diagonals and which are 'sides'?

Euclidean geometry, when it wishes to find truths about four-sided figures has recourse to making them parallelograms or rectangles, where special qualities are already put, artificially as it were, into them, by means of one's definitions. We see that projective geometry is much more general; the basic quality of the four-point belongs to all four-points, whatever their shape or size.

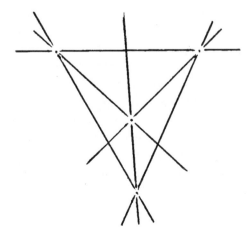

Figure 4.5

4.4 The harmonic property of the four-point

Any four-point contains six lines. These will be paired into three pairs, each pair consisting of lines which do not meet one another in points of the four-point. Take the common points of two such pairs and join them by a line. This line will be cut by the remaining pair in points which are harmonic with respect to these common points.

Exercise 4e

Dualize the construction of Figure 4.4. You will end with four harmonic lines in a point. Then dualize the above paragraph to find the harmonic quality of the four-line. Make a little sketch figure to illustrate.

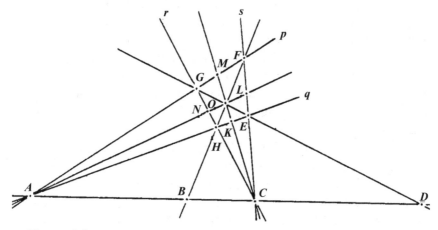

Figure 4.6

Now we will add two further lines to our harmonic figure, *AO* and *CO* (Figure 4.6).

We have already found that *EOGD* are harmonic. If we now project these four points from centre *C* we will find that *FMGA*, *LONA* and *EKHA* are three more harmonic sets of points. In fact, by similar projection methods we find that of the nine lines in this figure, each line contains just four points and each such set of four points is harmonic. The figure is saturated with the harmonic ratio.

This harmonic ratio takes a central place in projective geometry and we must try to come to a fuller understanding of its significance, although much must be left until a little later.

Exercise 4f

Draw four figures in each of which *A* and *C* are a certain fixed distance apart. Now let point *D*, in successive figures take up positions, (i) slightly to the right of *C*, (ii) as far to the right as the size of the page will allow, (iii) at infinity, and (iv) having passed through infinity, somewhat to the left of *A*. Now, completing the figures, find how the harmonic mate of *D*, point *B*, will move in sympathy with the movement of *D*. You will find that the two original points divide their line into two parts, internally from *A* to *C* and externally from *C*, through infinity to *A*. And the harmonic relationship creates a perfect correspondence between these two parts, in which the inner part becomes an inside-out picture of the outer, the centre point of the inner mirroring the point at infinity of the outer.

In this we see *one* of the very important features of the harmonic relationship. Up till now, our projective world picture, being non-metrical, has had no concept for the midpoint between two points. Now, without bringing any measurement to our aid, we can define the *midpoint* between any two points *X* and *Y* — it is the harmonic mate of the point at infinity of the line with respect to *X* and *Y*.

4.5 The absolute line

In pure projective geometry all lines are alike — no single one has importance more than another. We can, however, at a certain moment, single out one unique line of the plane and say, 'This shall be our *Absolute Line*. With regard to this we shall, as it were, take our bearings.'

Euclidean geometry, consciously or unconsciously, does this. It selects the line at infinity of the plane and makes that its absolute. Thereafter, all lines which meet at some particular point on the infinite line are said to be 'parallel.' Considered in this way, we approach the question of parallelism and the point at infinity from a purely projective aspect; in our first introductory chapter we approached the same matters from a more Euclidean aspect.

In projective geometry we are free to choose our absolute line wherever we wish. In Figure 4.6 for instance, we can choose line *ABCD* to be our absolute. We see immediately that *EFGH* becomes a parallelogram and *O* is the midpoint of *EG* and of *FH*, all with respect to line *ABCD* as the absolute. We realize that the old Euclidean statement, that the diagonals of a parallelogram bisect one another, is in fact a truth which applies, in its projective aspect, to all quadrilaterals whatever.

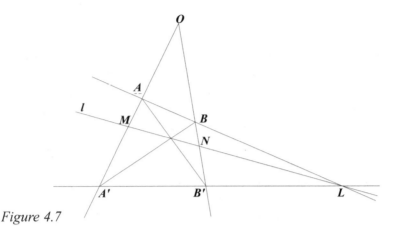

Figure 4.7

Now consider Figure 3.3 (page 39) concerning a homology.

We arrange that *A* transforms into *A'* and we let line *OAA'* meet the self-corresponding line *l* in *M* (Figure 4.7). We then find that *B* transforms into *B'*, according to the figure; and we let *OBB'* meet *l* in *N*. Now if we have arranged that *A* and *A'* are harmonic with respect to *O* and *M*, it can be seen, by projecting from point *L*, that *B* and *B'* must be harmonic with respect to *O* and *N*. We may say, that if in any homology one single point transforms into its harmonic with respect to the self-corresponding elements, then every other point of the plane does so too. Such a transformation is called a *harmonic homology*. By drawing *AB'* and *A'B* we see that *A'* transforms back into *A*; and similar for every point that is not self-corresponding. We are dealing with an *involution* (see Chapter 9).

Let us consider a harmonic homology with line *l* as self-corresponding line and point *O* at infinity in a direction at right angles to *l* (Figure 4.8).

We find that each point transforms into its symmetrically opposite point. Every case of symmetry is simply a case of harmonic homology with the self-corresponding point at infinity and the self-corresponding line as the axis of symmetry.

In a space homology, the self-corresponding line is replaced by a self-corresponding plane and, if it is a harmonic homology, it becomes, in fact, a case of mirror *reflection* with the self-corresponding plane as the surface of the mirror.

When we consider the symmetrical form of our bodies and of many of the organisms and creations of nature we see that this choice of an absolute by Euclidean geometry is not an arbitrary one. In many respects we are formed out of the infinite as our absolute.

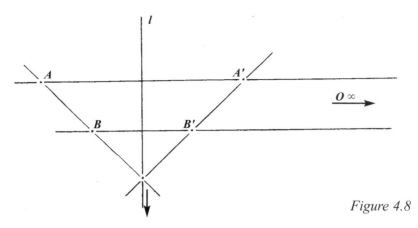

Figure 4.8

Exercise 4g

Returning to Figure 4.4 draw the upper half of this and add line *DH*. Where this meets line *p* join to *C*. You will have made a new little four-point consisting of *G, H* and two new points on lines *p* and *q*. By continuing this process you can make a series of dwindling four-points along lines *p* and *q* towards *A*. While doing this you will find that you have made a new set of intersections between *q* and *x* and that these are all collinear with *A*. Join them by a line and you will have a new set of little four-points. By continuing in this way, the whole page becomes covered with four-points — or as much of it as your time and the thickness of your pencil lead will allow; for the four-points dwindle and crowd in towards line *x* as though it were infinitely distant. In a way, for them, it is! Taken as the absolute, line *x* becomes the line at infinity for a perspective picture of congruent squares, or parallelograms, according to the viewpoint one decides to adopt, which cover the whole plane. This is called the *harmonic net*. The figure is full of the harmonic ratio and of one special *anharmonic* one. Every *three* consecutive points anywhere in the figure is collinear with either *A, B, C* or *D*, and is harmonic with it. But every *four* consecutive points in the figure, wherever they may be, with the single exception of *A, B, C* and *D*, have their cross ratio equal to $-\frac{1}{3}$. This is the cross ratio of four *equidistant* points and, in fact, each row of points in the figure is a perspective picture of equally spaced points on a line.

4.6 The diagonal triangle

Now consider again Figure 4.6. Our four-point *EFGH* has picked out for us nine lines. These lines are of two sorts. Six of them are 'sides' of the four-point and these six are grouped into pairs, each pair consisting of lines which do not meet in points of the four point. Each pair therefore meets in some point additional to the four points E, F, G and H. GF and EH meet in A, GH and EF meet in C, and EG and FH meet in O. The remaining three lines of the nine are those which join these extra points, A, C and O. The points A, C and O, with their common lines, are known as the *diagonal triangle* of the four-point.

Exercise 4h

Dualize the above paragraph to find the diagonal triangle of the four-line *pqrs*. You will find that the diagonal triangle consists of the lines AC, FH and GE. Of course, the whole of each line must be considered and the triangle, seen pointwise, is then OBD. Notice that these triangles are different from one another but that the points of each lie on the lines of the other.

Exercise 4i

There are two other four-lines which can be found from the six lines which are determined by the four-point $EFGH$. They consist of the lines p, q, FH, EG and r, s, FH, GE. Find their diagonal triangles. Note that in each case their points lie on the lines of the triangle ACO and the points of the latter lie on the lines of the former. But this relation does not exist between the diagonal triangles of the three four-lines.

Exercise 4j

Draw four red lines. They meet in six red points. Not every pair of points is connected by a line: you can draw three more blue lines. These meet in three blue points.

You can connect all the points so far with six additional green lines. These meet in four green points.

The picture now consists of thirteen points and thirteen lines. It is called the *thirteen-construction*.

Dualize by starting with four *green* points connected by six green lines, and notice that you get the same result. The picture is self dual.

Exercise 4k

Verify that the dimension theorem of Section 1.7 is self dual.

5. Correspondences

5.1 The algebraic background

In the preceding pages we discovered a number of remarkable relationships between the entities inhabiting space. These points, lines and planes began to reveal themselves as being close-knit into a set of inter-relationships of quite amazing precision, complexity and beauty. And at this point, it is useful for us to inquire just where the source of such wisdom is to be found. The answer, when we find it, may seem prosaic, even disappointing; but if we are to make the best progress possible in our studies, it is an answer which we should not try to avoid.

Our projective geometry is the wonderful and beautiful thing it is, because it is, at heart, *algebraic*. This book is planned to run its course without the use of algebra, as far as it is possible, but unless we at least acknowledge the basic source of the things we are studying, our understanding must be to some extent vitiated.

What do we mean when we say that a process, or transformation, is 'algebraic'? To answer that it means that it can be expressed by one or more of a series of algebraic equations, is to produce little more than a tautology. We must seek further, and deeper. These x's, y's, and z's, which litter the page of an algebraist — what *are* they in fact? Certainly not numbers, for the content of a number always possesses 'quantity,' and no one can attach a fixed quantity to any x or y. We must say that they are quantity-less entities which nevertheless behave, in all their inter-relationships, as though they were numbers. This is the wonderful thing about algebra. We regard these 'numbers,' but as soon as they are divested of their content, as soon as we can no longer be interested in what they *are*, all our attention becomes concentrated on what they *do* — on how they behave. In arithmetic we master the working of numbers; in algebra we study the laws of Number. It is not part of the purpose of this book to study the details of the algebra, but we do need to become aware of some of the basic things the algebraist has achieved. It is found that we can define a number of algebraic entities which we can name 'point,' 'line' and 'plane'; and it is found that if we impute to the interworking of our symbols just the working of the commutative, associative, distributive, identity and inverse laws of Number — nothing more — then these inter-relationships can be shown to run, exactly and in every way, parallel to those between the points, lines and planes of space, which we call the Propositions of Incidence. And we can come to see the wonder, beauty and precision of these propositions, and all that flows directly from them, as an outward, extensive expression in space, of the more inwardly-perceived world of the working of Number.

At this point it may be useful to put our considerations into somewhat of a historical perspective. There were times in the nineteenth century when our projective geometry came to be split into opposing camps; there were the Synthetists and the Analysts. The former wished to keep their geometry 'pure,' and unsullied by any hint of algebraic formalism or abstraction; they would not allow an algebraic symbol to appear on their page. The latter wished to keep their reasoning 'pure,' and undefiled by any reference to spatial intuition; if they could help it they would not allow a picture to appear on their page.

In due course it became realized how wrong-headed each of these attitudes was; the split was healed; and increasingly it became acknowledged that these points, lines and planes of space, together with the principles of incidence which link them, are true, valid and exact spatial representatives of the algebraic relationships which lie behind them.

Now we saw already that in our geometry we are very interested in establishing correspondences between sets of elements, perhaps the points of two lines, or the lines of two points, or even between the points of a single line, or the lines of a single point. The most important kind of correspondence is the simplest, where each element of the one set corresponds projectively to just one of the other, and vice versa. This is known as the 1–1 (one-to-one) projective correspondence or projectivity (see Section 4.1). There are, for instance, various ways in which it can be arranged geometrically that, say, the points of two lines correspond in such a way.

But the algebraist has discovered a way of establishing a one-to-one correspondence between the relating symbols of his number system. He has found a little equation which enshrines *all* the possibilities of these one-to-one correspondence between two numbers, x and x'. It is very simple, and here it is:

$$axx' + bx + cx' + d = 0$$

From this we can immediately deduce some important facts.

— Firstly, because there are just three independent coefficients, the complete correspondence is determined by the giving of any three pairs in the correspondence.

— Secondly, if the elements all belong to the same base (the correspondence is *co-basal*) we can ask whether any elements will correspond to themselves. To find this, we put $x = x'$ and the equation reduces to:

$$ax^2 + (b + c) x + d = 0$$

This is an ordinary quadratic equation — and, as is well-known, it must have just *two* solutions, which however may be real, or imaginary (conjugate complex), or, in a special case, real and co-incident.

— Thirdly, by simple algebraic manipulation it is easy to show that the cross ratio of any four elements must be equal to that of their corresponding elements.

When a correspondence is of the form $axx' + bx + cx' + d = 0$ we call it a *linear* one-to-one correspondence, and it assumes a position of very special importance in our geometry. These linear correspondences in algebra match exactly the projectivities in our geometry.

Now we can say from what has gone before, that if we establish such a correspondence between these algebraic entities, using only these qualities of incidence, we can expect that it stays wholly within the algebraic realm. The criteria for achieving this, made sufficiently precise, and treated rigorously, would provide total proof that the resulting correspondence must be linear.

In this book we shall content ourselves with stating the criteria rather more descriptively. We shall say that if we work, using only the principles of incidence between points, lines and planes, then the resulting correpondence is indeed linear, and therefore projective. But in doing this we must be sure that our correpondence is universally valid; that is to say that, applied consistently, it must be true for every element whatever of the forms we are working with.

Now if the reader has found some of the foregoing difficult to understand, he should not let this worry him unduly. In order to follow what comes later in the book he needs to remember these four important factors:

1) that a projective one-to-one correspondence is completely fixed by the giving of three pairs of corresponding elements (note: this is true for what are called one-dimensional forms, for instance, the points on a line or on two lines, or the lines of a point; when it comes to correspondences embracing the elements of a whole plane, we shall work out similar but rather more complicated rules);

2) that if it is a correspondence between elements of the same form (say the points of a single line, or the lines of a single point) then there must be always just *two* self-corresponding elements, although these may sometimes be imaginary, and therefore, invisible on our figure;

3) that the cross ratio of any four elements will always be the same as that of their four corresponding elements;

4) that if we make a correspondence between the elements of our forms, using only join and meet, ensuring that it is universally one-to-one (that is, that it embraces every element whatever of our field) then it will be a projective correspondence with all the qualities which are described above.

In particular, it is from this background that the whole concept of imaginary points and lines arises; and it is these which we are going to study, later in this book, largely without algebra.

5.2 Perspectivity

Let us apply these thoughts to Figure 5.1. Can we say, by the previous considerations that the ranges of points *ABCD*... and *A'B'C'D'*... are in projective one-to-one correspondence? Firstly we notice that point *O* is outside lines *x* and *x'*, so that, by the conditions of the

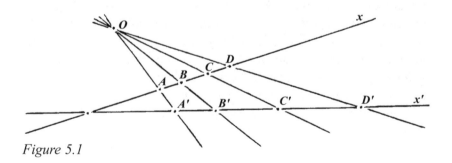

Figure 5.1

problem, O must be distinct from any of the points in those lines. Take any point A of x; A and O must have just one, and only one, line in common. This line and line x' must have one and only one, point in common — the point A'. This makes it clear that any point of x is related uniquely to just one point of x'. Now consider point A' of x'. A' and O have just one line, and only one, in common; this line and line x have just one, and only one, point in common — point A. Thus, it is also clear that any point of line x' is related uniquely to just one point of x. We have established by quite exact thinking that these two ranges of points are in projective one-to-one relationship; and by the parallel with the algebra, if for no other reason, we might have expected that cross ratios would be invariant from one range to the other.

In Section 4.1 we found quite empirically, that this is the case. When we tried it, we found that cross ratios did indeed come out the same from one line to the other; and we have since assumed that this is indeed a quality of the projective process, without having any further proof of it. Now, we begin to see a little more deeply into the considerations which lie behind this truth.

So we can say now that all perspectivities are projective one-to-one correspondences, and from this it follows that all projectivities also have this quality.

Next we must note two important qualities which are peculiar to perspectivities.

Firstly, we must ask ourselves how many pairs of points we can relate to one another arbitrarily and be sure that such pairs do, in fact, belong to a perspectivity. Having decided that the ranges on lines x and x' are to be in perspective, we can surely put down two pairs of points, A and A', B and B'. Now we find that point O is determined, for it must be the common point of lines AA' and BB'. If we now put down a new point C on x, we are no longer free to put C' anywhere we like on x'. Its position is clearly already determined; it must be the common point of lines OC and x'. Thus we may state our first law of perspectives thus:

> *Any perspectivity is completely determined when two pairs of corresponding points are given.*

The other important fact to notice is that the common point of the lines x and x' is obviously self-corresponding.

The converse of this is also necessarily true: if two ranges in projective one-to-one correspondence on two lines have the common point of those two lines self-corresponding, the relationship is a perspectivity and it follows that the common lines of corresponding points are all concurrent. This result is not so obvious, but it follows from the fundamental theorem to be discussed later.

This simple fact, and its dual, can often be used to bring us exact information about various problems.

For instance, consider Exercise 2b, Section 2.7 (page 30), the theorem of Pappos. We produce the figure here, with two further points lettered, *S* and *T* and the common point of *x* and *x'* lettered *K* (Figure 5.2).

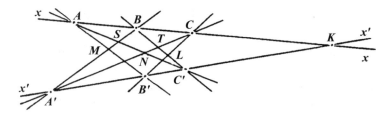

Figure 5.2

Now we can find a projective one-to-one relationship between the points of lines *x'* and *A'B* by projecting the one onto the other from centre *A*. We then find that the points *KC'B'A'* correspond to points *BSMA'*.

Similarly, by projecting from centre *C* we have a projective one-to-one relationship between *x'* and *BC'* in which we find that points *KC'B'A'* correspond to points *BC'LT*.

Now it is clear that if lines *BA'* and *BC'* are both in projective one-to-one-correspondence with line *x'* they are also in this relationship one with another, and we have it that points *BSMA'* correspond, in that order, with points *BC'LT*.

We see then that point *B* is self-corresponding in the two ranges; therefore the relationship is a perspective and the joins of corresponding points will all be concurrent at some point. We can tell easily what this point is, because *C'S* are corresponding points, as also are *A'T*. Therefore *N* is the centre of the perspective and it follows that the line joining the remaining pair *LM* must pass through *N*.

Now we have proved that *L*, *M* and *N* must lie on one line.

Exercise 5a

The same fact can often be used to prove the concurrency of a number of lines. For instance, consider the figure of Exercise 2c, before dualizing. Prove that the three lines BB'' CC'' and $B'C'$ must meet in one point. (Hint: project the points of the line BC on to lines AB and AC from centres C'' and B'' respectively. It would be a help to letter the point at which BC meets the line through A).

Exercise 5b

ABC is a triangle and l is any fixed line passing through C. P is a variable point on l. AP meets BC in Q and BP meets AC in R. Prove that the line QR, as P moves, always passes through a fixed point of AB.

The dual of these facts about perspective ranges is so important that we include it here with a Figure (5.3).

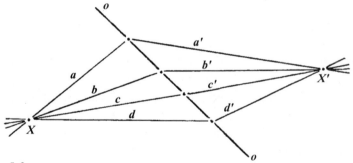

Figure 5.3

Two pencils of lines (through centres X and X') are said to be *in perspective* when the meets of corresponding lines all lie on a line (line o). We notice that the common line of the two pencils is self-corresponding.

Conversely, if two pencils are in a projective one-to-one relationship in such a way that their common line is self-corresponding, then the pencils are perspective and the meets of corresponding lines are collinear.

Exercise 5c

Dualize the figures and proofs of Exercises 5a and 5b.

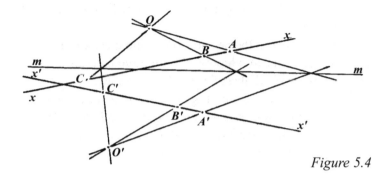

Figure 5.4

5.3 Projectivity and the fundamental theorem

Let us suppose that we establish a perspectivity between the points of line *x* and line *m* and another perspectivity between the points of *m* and a line *x'* (Figure 5.4).

We project the range of points *ABC...* of *x* from point *O* on to line *m* and then project the resultant range from *m* to line *x'*, using centre of projection *O'*. Clearly we have established a projective one-to-one relation between the points of *x* and of *x'*. The cross ratio of any four points of *x* will be the same as that of their corresponding points on *x'*. We can now ignore line *m* and say that we have a projective one-to-one relation between the points of *x* and *x'* and any four points of the one will have the same cross ratio as the corresponding points of the other. But we shall not find that the joins of corresponding points, *AA'*, *BB'*, *CC'* etc. all pass through one point.

Neither will we find that the common point of *x* and *x'* is a self-corresponding one. A glance at the figure will show us that this point must be considered twice. We can consider it as a point *K* of *x*. So considered, its corresponding point is *K'* of *x'*. Also, we can consider it as point *L'* of *x'*. Its corresponding point is then *L* of *x*.

Clearly a *perspectivity* is a special case of a *projectivity* and we can recognize it by the fact that the common point of the two lines is self-corresponding.

We now have to answer a question which is crucial. On two lines *x* and *x'*, how many pairs of points *AA'*, *BB'*, etc. can we put, quite arbitrarily, and be sure that:

— they will all be pairs in some projectivity, and
— that this projectivity will be unique to these pairs, that is to say, that there is no other projectivity which would unite all these pairs.

In other words, how many pairs of points will completely determine a projectivity?

We will deal with the second part of the question first. Suppose we have a projectivity in which we know three pairs of points, *AA'*, *BB'*, *CC'*. If we now put down a fourth point *D* anywhere on *x*, we will have a definite cross ratio of the four points *ABCD* and we know that this must be the same for the cross ratio of the four points *A'B'C'D'*. But we know (Section 4.1) that given *A'B'C'* and a definite cross ratio for the four points, there is only one position which *D'* can assume. Therefore, it follows that given *AA'*, *BB'* and *CC'* we have no freedom to assert that any other pair *D'D'* we like is also part of the projectivity. It follows that the number we are looking for must be less than four and that three pairs of points, if they belong to the projectivity, will completely determine it.

It remains to ask, 'If we have three pairs of points, arbitrarily chosen, are they bound to form part of a projectivity?'

We will now show that the answer is *yes*. If we put three pairs of points on two lines, placed arbitrarily, and we find a general method of linking them in projectivity, the question is answered.

Place points *ABC* and *A'B'C'* on *x* and *x'* respectively (Figure 5.5). Through *A'* draw any intermediate line *m* and take any centre

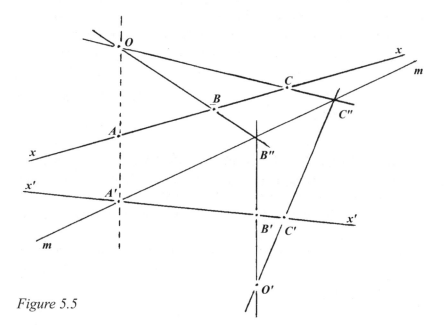

Figure 5.5

of projection O which lies on the line AA'. From O, points ABC will project on to m as points $A'B''C''$. Let O' be the meeting point of lines $B''B'$ and $C''C'$. Now $A'B''C''$ project from O' into $A'B'C'$. The projectivity is established!

This is so fundamental a law that it is often known as the *Fundamental Theorem:*

> *Any three pairs of points will establish, and completely*
> *determine, a projectivity.*

It is obvious that if our statement (Section 4.1) about the invariance of cross ratios under projection has any validity at all, then had we taken any other line m through A', or any other point O on the line AA', with consequent alteration in the position of O' and if we had taken any fourth point D of x and had found D' of x', firstly before altering m and/or O, and secondly after, then in each case the same point D' must result. The whole structure of projective geometry, one might almost say of space, rests on the fact that it is bound to be so. It is a fundamental truth[7].

Exercise 5d

Try it.

Exercise 5e

Take any two points X and X'. Put three pairs of lines a, b, c and a', b', c' respectively through them. Now by dualizing the above construction establish a projectivity between them. Find the line d' corresponding to any line d of X.

5.4 Co-basal projectivities

A very important case arises when we project a line, point by point, one-to-one, back into itself. This is sometimes referred to as transforming the line into itself; you apply the projectivity, or transformation, and each point transforms into its corresponding point along the line.

A very simple construction will illustrate such a transformation (Figure 5.6).

We have a line x which we will transform. We put down an intermediate line m and two centres of projection O and O'. We take any point

---

* See also Exercise 4c, page 46.

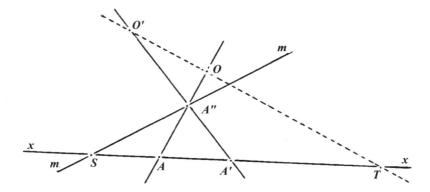

Figure 5.6

A of *x* and project it from centre *O* into *A"* on line *m*. We then project this
point back onto *x* from centre *O'*. The line is projected out of itself and
then back again. This forms a projectivity between *co-basal* ranges; it is
exceedingly important and we shall have much more to do with it. At the
moment we have just one crucial question to ask. Will there be any
points which project back into themselves — and how many?

Suppose that three points *A*, *B* and *C* project back into themselves.
If we put down any fourth point *D* it will, with the other three, have a
definite cross ratio and the four points *A'B'C'D'* must also have this
cross ratio. But as *A*, *B* and *C* are unaltered it follows that *D* must be
so too. In other words, if three points are self-corresponding all points
of the line will be also — and there will be no transformation.

Therefore, our number must be less than three.

Now by looking at our figure we can find two self-corresponding
points. They are *S* (where *m* meets *x*) and *T* (where *OO'* meets *x*).

It can, in fact, be shown algebraically that

> *every co-basal-projectivity has just two self-corresponding
> entities,*

as was described at the beginning of this chapter. They may be real or
imaginary and, if real, they may, in special cases, be co-incident.

This is a fact of fundamental importance.

We have already met such a transformation of a line into itself.
Consider the homology of Section 3.3, page 39. Each line through the
centre of homology *O* is transformed into itself in just this way. The
self-corresponding points are point *O* itself and the common point of
the line with line *l*.

6. The Conic

6.1 Construction of a conic

In Exercise 5e (page 64), we found that we could pair the lines a b c and a' b' c', passing through centres X and X' respectively, in such a way that they correspond in a projectivity. This projectivity would then link every line of X with its corresponding line of X', one-to-one. If the common line of the pencils is self-corresponding; that is, if XX' corresponds to $X'X$, then the relationship would be a perspectivity and meets of corresponding lines would lie on a line. In general, however, this will not be the case, and we have now to investigate where these meets of corresponding lines normally lie.

If the meets of corresponding lines do not lie on a line they will lie on some sort of curve. We start by constructing such a curve, which will appear to be a *conic*. The most convenient way to form the projectivity between the pencils in X and X' is by a double perspective, the way we did in Section 5.3.

We put down the centres of our two pencils, X and X' and two lines o and o' (Figure 6.1). A line a, through X, projects on to line o in point A. This is projected from intermediate point M into point A' of line o' and a' is the line from A' to X'. a and a' are now two corresponding lines in the two pencils and their common point P is a point of the curve. As the line, a, turns about point X it generates a range, $AB...$ of points on line o. These are projected into the range $A'B'...$ of line o' from point M and thence into the pencil of lines $a'b'...$ through X'. The two pencils are thus in a projective one-to-one correspondence. As this construction will be used repeatedly in what follows, it is important that the reader carry it out by doing the following exercise.

Exercise 6a

Make this construction using as many lines through X as convenient and possible.

It will be found that only a part of the curve can be constructed because so many of the lines through X meet line o somewhere off the page. However, there are certain considerations which can help us. Consider first that line of X which passes through M. It may meet

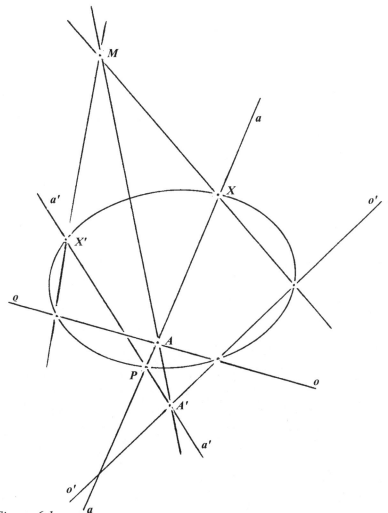

Figure 6.1

line *o* off the page but this does not hinder us, for the line from this point to *M* is already the line through *X* with which we are dealing. Quite obviously the point in which this line meets line *o'* is a point of the curve. So we can mark in as points of the curve straightaway the points where the lines *MX* and *MX'* meet lines *O'* and *o* respectively. Also we find that the curve always goes through the common point of *o* and *o'*.

Now consider the line of *X* which passes through *X'*. This will correspond to some line of *X'* and, of course, must meet it there. So

we see that the curve must pass through the centres of the two pencils, X and X'.

We cannot cope with a line through X which meets o at some distant
point, but we can with that line which meets o at infinity. This and the
point given by the line of X' which meets o' at infinity can both be constructed by following the method for all ordinary lines of X and X', step
by step.

It is a good thing to make the construction using a similar layout of
points to that which we give here and again with some quite different
arrangement.

6.2 Order of a conic

Now we must see how much we can learn about this curve. Firstly, let
us consider again how it has been made.

The pencil through X, *abcd*... is projective with the pencil through
X', *a'b'c'd'*... . Now cut the curve by any line. This projectivity will be
transferred to the points of the line, A corresponding to A', B to B', etc,

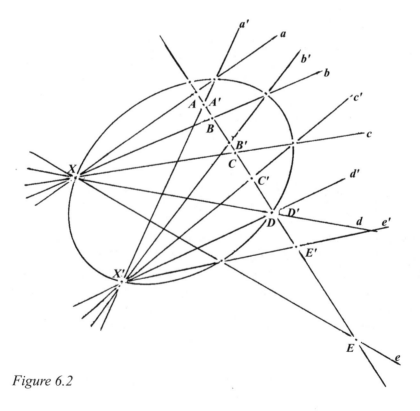

Figure 6.2

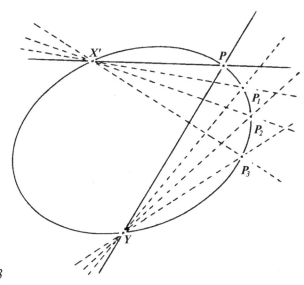

Figure 6.3

as shown on Figure 6.2. A moment's thought will serve to show that where the curve meets the line there must be a self-corresponding point of the projectivity, note DD' in our figure. But this is a co-basal projectivity and we have seen (Section 5.4) that such a projectivity always has just two self-corresponding points, real or imaginary. Therefore, any line meets this curve in just two points. It is a curve of second order.

The curve is called a *conic* because it can be shown that it is identical with the *conic section* of classical geometry. We shall use the name here but will not assume any of the qualities ordinarily associated with it until we have proved them from our own studies.

6.3 Projective pencils on a conic

Now let us consider the curve again. It is made up of the common points of corresponding lines in two projective pencils through X and X'. Let us take any third point of the curve, Y and through Y any line we like (Figure 6.3).

Since the curve is of second order we know that this line must meet the curve in one, and only one, other point, which we will call P. It is clear that as the line YP turns about Y, the curve is establishing a one-to-one correspondence between the lines centered in Y and those in X'. We notice that this correspondence derives immediately, and only, from the fact that the curve has been proved to be of second order, which fact follows solely from the algebraic relationships which have formed the curve (see the little quadratic equation of Section 5.1. No other considerations have entered the argument; so we can say, considering fact 4 of Section 5.1 (page 58), that this correspondence is a projective one of the kind described in that section. Thus we see that the points X and X' are in no way special points of the curve. Exactly the same curve could have been constructed from pencils through X and Y, or through any two other points of the curve.

We see that the above reasoning follows simply from the fact that the curve is of second order, and this allows us to state a theorem of great importance.

> *Every curve of second order is a conic*

(as described in Section 6.1 and 6.2). There are no other curves of second order. This is a fundamental truth of our projective geometry.*

6.4 Cross ratio on a conic

Now let P_1, P_2 and P_3 be three further points to which P moves as YP revolves around Y (Figure 6.3). It is clear that the cross ratio of the lines YP, YP_1, YP_2, and YP_3 must be the same as that of the lines $X'P$, $X'P_1$, $X'P_2$ and $X'P_3$ and, of course, of the lines XP, XP_1, XP_2 and XP_3. In fact, we can join the points P, P_1, P_2 and P_3 to any other point of the conic and the resulting pencil of four lines will have the same cross ratio.

This is a fundamental quality of the conic. In fact, it would be possible to define the curve in this way. Take any four points and join them to a fifth. Now let the fifth move in such a way that the cross ratio of the lines joining it to the other four remains constant. Its locus will be a conic.

If then, four points of a conic can be joined to any variable point of the conic, and will always have the same cross ratio wherever that point may move, we are justified in speaking of this cross ratio as

---

* For a closely-reasoned mathematical proof we need more than we have developed so far.

belonging to those four points, considered as members of this particular conic. (It would be possible to draw quite a different conic
through those four points, and then they would have a different cross
ratio considered as members of *that* conic.)

It now becomes clear that in some respects the conic has a similar
quality to that of the line; points on it can be determined by numbers,
and, indeed, any relationship which points assume on a line can be
duplicated on a conic. For instance, we could imagine two ranges of
points on a conic which would be projective. The cross ratio of any
four points would be equal to that of their four corresponding points.
This would, in fact, be a co-basal projectivity, and we shall find that
all the truths we have found about such projectivities on a line will
be valid for projectivities on a conic.

6.5 Projectivity on a conic

Let us construct such a projectivity in the simplest way possible. We
take a conic and any point *O* not on the curve (Figure 6.4).

Now consider any point *A* of the curve. *O* and *A* must determine
one, and only one, line. This line, meeting the curve at *A*, must meet
the curve at one, and only one, other point *A'* (because it is a second

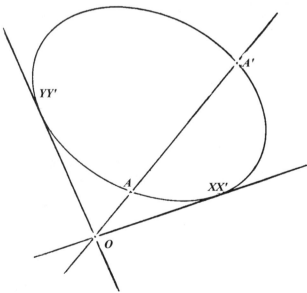

Figure 6.4

order curve). Thus, there is a one-to-one correspondence between
the ranges A and A' as the line revolves around O. This projectivity,
like all others, must have just two self-corresponding points, $X=X'$
and $Y=Y'$ on our figure. Clearly these are the points at which the tangents from O touch the curve.

From this it follows that if a curve is of second order it must also
be of second class, for every point of the plane will send just two
tangents to it. (A corresponding rule by no means holds for higher
order curves.) It is now clear that if we take a pointwise conic and
draw all its tangents we shall get a curve of exactly the same nature
as if we dualize the construction with which we started this chapter.
This is what we mean when we say that the conic is a self-dual
curve. If we polarize it we do not, in general, get identically the
same curve, but we always get one with identical properties; that is,
another conic.

Exercise 6b

Dualize the whole of this chapter so far. The dual of Figure 6.1 is very
important and is left to the student. The dual of Figure 6.3 should also
be worked out independently by the student, but is so important that it
is included here for reference (Figure 6.5).

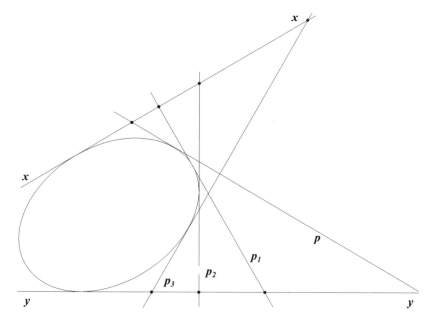

Figure 6.5

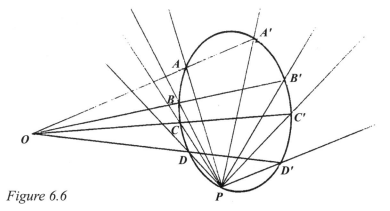

Figure 6.6

The same construction leads to another remarkable fact. We can take any four lines through point *O*, to meet the conic in the eight points *A B C D* and *A' B' C' D'* (Figure 6.6).

If we now take any ninth point, *P*, of the curve, and join it to *A B C D* on the one hand, and to *A' B' C' D'* on the other, these two sets of lines will have the same cross ratio; and, of course, as *P* moves around the curve this cross ratio will remain unchanged. (Note: this is *not* the same cross ratio as that of the four lines through *O*.)

It is a useful exercise to draw this figure accurately and check the equality of the cross ratios. It can be done easily by measuring the spacing of the eight points along one of the lines through *O*.

But we can go further. The pencil of lines through *P* is a remarkable one. There was nothing in our construction of Figure 6.6 to say which way the meeting points were to be labelled. We could have had it like in Figure 6.7.

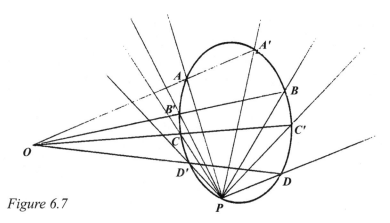

Figure 6.7

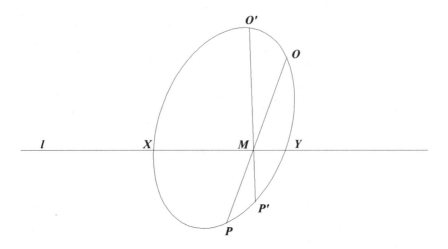

Figure 6.8

This gives us two new sets of four lines *PA PB PC PD* and *PA'*
PB' PC' PD'. These will give a different cross ratio from the previous sets, but the cross ratio of these two sets will again be equal to
one another. In fact, there are eight different ways in which the lines
of *P* can be 'scrambled' into pairs of sets of four, and in each case
the members of the pair will have equal cross ratios.

In Figure 6.8, *O* and *O'* are any two fixed points of a conic and *l*
any line. Let *OP* meet *l* in a point *M* and let *O'M* meet the conic in
P'. As *P* moves around the curve, *P'* moves in one-to-one correspondence with it. This is perhaps the easiest way to construct a general one-to-one projectivity on a conic. The double points of this
correspondence are at *X* and *Y* where the line *l* meets the curve.
Clearly this gives us easy power to have our double points real,
imaginary, or co-incident, according as to whether we let the line *l*
meet the curve in real points, or not, or we make it tangent to the
curve.

6.6 Degeneration

Let us now construct a projectivity between two pencils in rather a different way. Draw two pencils of lines, equiangular, with, say, five degrees between each line. Name the common line of the two pencils $a=a'$ and then name the lines of one pencil consecutively $bcde...$ in a clockwise direction and the lines of the other $b'c'd'e'...$ in an anti-clockwise direction. It is clear that the cross ratio of any four consecutive lines of the one pencil must be equal to that of any consecutive four of the other. Thus we can make any line of the one correspond to any line of the other and then all the rest will correspond in order.

First let us make line d correspond with, say, m'. Then c will correspond with l' and e with n' etc. The common points of corresponding lines will give us a conic section.

Now draw on the same figure, the curves which result from making d correspond first with j', then with g' and then with e'. It will be seen that these curves grow progressively straighter during most of their length. The final one is almost completely straight for most of its way, with two sharp twists near the centre.

Now let d correspond with d'. This, of course, will mean that a corresponds with a'. The projectivity has become a perspectivity and meets of corresponding lines lie on a line (Section 5.2). Obviously, there is a continuous metamorphosis from the curves into this straight line. But the law of the conic is that every point which is common to corresponding lines is a point of the conic. Clearly all points of the self-corresponding line $a=a'$ obey this law, as well as the meeting points of the lines d and d', e and e', f and f' .... etc.

We say that the conics degenerate into this pair of lines — the line $a=a'$ and the line of points dd', ee', $ff'...$ etc. and we can refer to this pair of lines as a degenerate conic.

We shall, in fact, find that we can regard any pair of lines as a degenerate conic and that many of the laws of ordinary conics will apply also to such a pair.

By now making d correspond with c' we can make the conic reform itself on the other side, as it were, of the degeneration.

Exercise 6c

By lettering both pencils in the same direction, say, both of them clockwise, the curves which we should have gotten would have been circles and we should have had an interesting illustration of the Euclidian theorem about angles in a segment of a circle. Into what two lines would these conics have then degenerated?

Exercise 6d

Draw an equiangular pencil of lines — say every five degrees — and cut them by two lines, x and x', which meet on one of the lines of the pencil. Now letter the points of x consecutively A B C D... and the points of x' A' B' C' D'..., letting the common point of the two lines be $A = A'$. Clearly again, any four consecutive points of the one line can be related projectively to four consecutive ones of the other.

Draw the two linewise conics which arise, from relating firstly, A to B' and secondly, A' to B. We see two conics lying on either side of the degeneration. In between them the degenerate conic is formed of two pencils of lines, one of which is the original pencil with which we established the projectivities and the other is the pencil of all lines through the point $A = A'$.

Notice that when a pointwise conic degenerates, the order always remains — the line-pair can still be cut by any line of the plane in just two points, but we can no longer speak of class with respect to it. Similarly, when a linewise conic degenerates, it remains second class — every point of the plane contains two lines of it — but one can no longer speak of its order.

In our alternative definition of a conic, 'take any four points and join them to a fifth ... etc.' (Section 6.4), we would have had to specify that no three points of our five must lie in a line, if we wanted to ensure that our conic would not be a degenerate one. Seeing that we are willing to consider degenerate conics in our scheme of things there was no need to make this proviso.

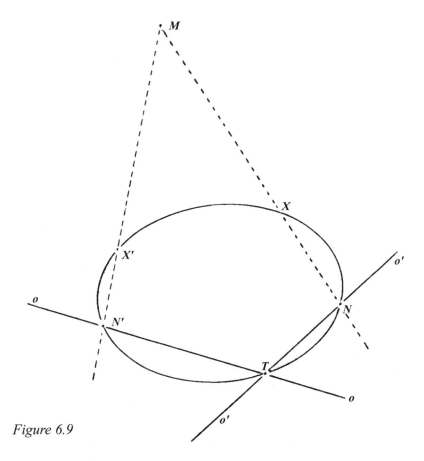

Figure 6.9

6.7 Tangents to a conic

We repeat here the figure of Section 6.1, by which we can construct a pointwise conic (Figure 6.9).

X and X' are the centres of the projective pencils, the lines of X projecting to o and those of X' projecting to o', the two ranges being connected by lines through the intermediate point M. It will be remembered that we found that the curve is bound to pass

— through T, the common point of the lines o and o',

— through the centres of the pencils, X and X'

— and also through the points N and N', where the lines MX and MX' meet o' and o respectively.

Now consider any line a of X. It will meet its corresponding line a' of X' in point P of the curve. The conic, being a second order curve, must meet a' in just two points and these are obviously X' and P. Now suppose P to move gradually towards X'. In the moment when P and X' coincide, the line a of X will pass through X' and its corresponding line a' will be the tangent to the curve through X'.

We thus see the rule:

> *the line of either pencil which passes through the centre of the other pencil has for its corresponding line the tangent through the centre of the other pencil.*

6.8 Determining a conic

Now we must ask ourselves, how many points are needed to determine a conic? In other words, what is the greatest number of points which I can put arbitrarily on to any plane and know that there is bound to be one, and only one, conic which goes through them all?

It is clear that once I have put down two projective pencils the conic which is going to come from the meets of corresponding lines is already fixed whether I have actually drawn it or not. Now we know by the Fundamental Theorem (Section 5.3) that to determine a projectivity between two pencils, or ranges, I must, and can only, choose three arbitrary pairs. The meeting points of these fix three points of the conic; but in doing this I must, of course, have already chosen centres for the pencils and we know that the conic will also go through these. Therefore, I have chosen five arbitrary points and I know that I can construct just one conic through them.

> *A conic is determined by any five arbitrary points in a plane.*

If any three of these points are collinear the conic will be a degenerate one.

By dualizing the above we arrive at the result:

> *A conic is determined by any five arbitrary lines in a plane as its tangents.*

How can we construct the conic which passes through five given points?

Let the points be A, B, C, D and E. Choose any two of these, say A and B, as centres of the pencils. Join AC, AD, AE and BC, BD, BE.

Now construct a projectivity making AC correspond to BC, AD to BD and AE to BE (Section 5.3). Meets of all corresponding lines will now give the required conic.

It is often easier to attain the same result by considering Figure 6.9. It is clear that as soon as we have fixed X and X', o and o' and M, the conic is determined — all further points follow automatically. But in fixing these things we have already fixed five points — X, X', N, N' and T. Here are our five points again! Therefore, given five points, we can choose any four to play the parts of $X\,X'\,N\,N'$. By suitably joining them we can find the position of M. And by joining N and N' to the fifth point, which is to play the part of T, we immediately find lines o and o'.

Exercise 6e

Put down five points, no three collinear in the most unlikely-looking combination possible, and draw the conic which passes through them all.

Exercise 6f

Dualize. Draw the conic which is tangent to five unlikely-looking lines.

Notice that it is possible to carry out the construction with line o passing through X and/or with line o' passing through X'. In such a case, N and X become coincident and obviously, the line MXN becomes a tangent to the curve (as will line $MX'N'$ with N' and X' co-incident).

Exercise 6g

Construct a conic through three given points so as to be tangent to given lines through two of them.

It is obvious that if five points determine a unique conic, the greatest number of points in which two different conics can meet is four. In an algebraic treatment, these points are given by the four roots of two simultaneous quadratic equations. If imaginary numbers are taken into account it is found that these four roots can be found between any two equations whatever. In our geometry we will admit the existence of imaginary points and lines and we will say that any two conics whatever meet in just four points, real or imaginary.

6.9 The three kinds of conic

Projectively speaking, all conics are equivalent to one another. But because of the way in which our consciousness is set into the world, because we find, as it were, a natural absolute in the line at infinity, conics seem to us to be of three distinct types, according to the relation they have to the line at infinity.

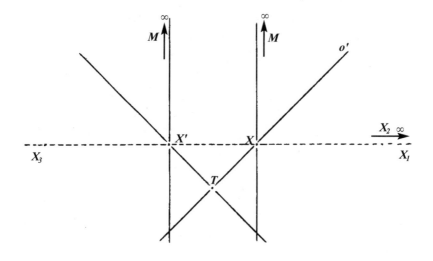

Figure 6.10

Construct a conic in the following way (Figure 6.10). Let the intermediate point M be at infinity, in a direction at right angles to the line XX'. Let o' and o pass through X and X' respectively, at 45° to the lines XM and $X'M$. Obviously, the conic will be tangent to the lines XM and $X'M$. By considering the line of X which is parallel to o, it will become clear that the conic must pass through a point which is the reflection of T in the line XX'. A few moments Euclidean thought will serve to show what the curve must be! But it is a good thing to construct it by the projective method.

Now on a new figure, move X to X_1, on the right hand side of the page. Keep M, X' and T fixed and keep o and o' still through X' and X. Construct the new curve. This new curve does not meet the line at infinity in real points; it is called an *ellipse*.

Now move X to X_2 which is the point at infinity of line XX'. Repeat the construction. Clearly this new conic will not only reach to infinity, it will be exactly tangent to the line at infinity. For X and M both being at infinity, their common line must be the line at infinity. So, as X moves outwards a moment will come when the conic is tangent to the line at infinity. Then it is called a *parabola*.

Next we let X pass right through infinity and come to X_3 on the left side of of the page. Keep all the other elements of the figure fixed as before and construct the new conic. You will see that the other end of the ellipse which now passes through X_3, has been taken, as it were, through infinity and has come back 'from the other side.' Any conic which meets the line at infinity in two real points is called an *hyperbola*.

A number of interesting constructions can be tried. For instance, repeat the last figure now putting X and X' at infinity in positions at right angles to one another and let M be in the centre of the page. Let o' and o pass through X and X' respectively and pass 2.5 cm above and 2.5 cm to the right respectively of M.

Exercise 6a

Do this.

This curve will prove to be a hyperbola; it is bound to pass twice through the line at infinity in the real points X and X'. The lines MX and MX' are tangents to the curve and they clearly touch it at the points in which it meets the line at infinity. Any tangent to a curve which touches it at infinity is known as an *asymptote*. If a conic has its asymptotes at right angles it is called a *rectangular hyperbola*.

Notice carefully the way in which a point of the curve is obtained. We draw any line through M and let it meet o in S and o' in S' (Figure 6.11). The perpendiculars through S and S' meet in point P of the curve. A few moments consideration of the Euclidean intercept theorems will show that if S' is two units horizontally to the right of M, then S must be one-half unit vertically above it, if S' is three units to the right, S must be one-third above; and so on. In fact, the distance of S vertically above M is always the reciprocal of the distance which s' is horizontally to the right of it.

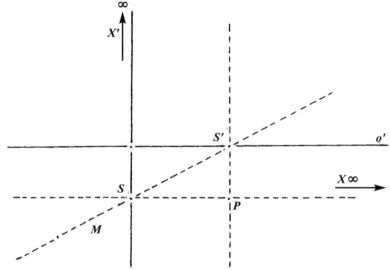

Figure 6.11

Clearly, if we take M to be the origin of a Cartesian system of co-ordinates this curve will have the equation

$$y = \frac{1}{x}$$

This is the well-known Cartesian equation for a rectangular hyperbola and we get our first definite indication that the conic of projective geometry is indeed the same curve as that which is dealt with in the analytical methods.*

6.10 Theorems of Pascal and Brianchon

One of the most famous theorems about conics is that discovered by Blaise Pascal in the seventeenth century. It concerns any six points on a conic. The six lines joining them form a hexagon inscribed to the conic. Number the sides 1 2 3 1 2 3 in order. Then any two sides with the same number are called opposite sides. The theorem states that, no matter how irregular the hexagon may be, the common points of the three pairs of opposite sides lie on a line (Figure 6.12).

---

* One has to take the distance from M to o and o' as unit.

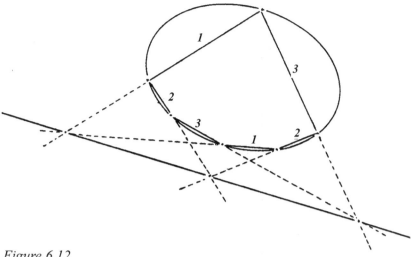

Figure 6.12

We notice that neither the shape of the hexagon, nor the lengths of its sides, have any bearing on the problem. The six points lie on a conic and the three points lie on a line; the essential facts are those concerning incidence; it is a truly projective problem.

This being so, it does not matter in which order we join up the points; the essential conditions of incidence are still fulfilled; the points still lie *on* a conic. This means that the hexagon may be re-entrant just as well as not. Given six points on a conic there are, in fact, no less than 60 different hexagons which may be made from them, according to the order in which we decide to join the points (6!/12).

Let us try making a re-entrant hexagon from six points on a conic (Figure 6.13).

This is just one of the 60 possible hexagons. For this one we should have to take the points in the following order — *AECFBD*. (Note. In making a re-entrant hexagon in this way it is important to see that there are just two lines of the figure going to each of the six points — not one to one of them and three to another.)

Now let us consider the four points *F, E, D* and *B*. These will have a definite cross ratio which will appear in any pencil formed by joining them to another point of the curve. Thus we can project these four points into the line *FB* from centre *A*. We see then that *FEBD* projects into *FLSB*. Similarly, by projecting from centre *C* into line *DB* we see that *FEDB* projects into *TNDB*. Putting these two results together we

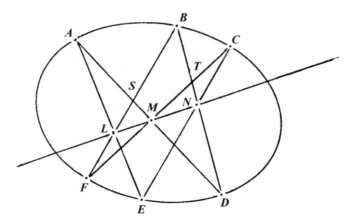

Figure 6.13

find that we have a projectivity in which *FLSB* corresponds with
TNDB.

We see that *B* is a self-corresponding point and, therefore, the
projectivity is a perspectivity and joins of corresponding points
must be concurrent.

Now we see from the figure that *FT* and *DS* meet in point *M*.
And from the proof above, it is clear that *LN* must also pass
through this point. In fact, we have proved that *L*, *M* and *N* must
lie on a line.

It is instructive to note that this proof is almost identical with
that which we gave for Pappos' Theorem (Section 5.2). In fact,
Pappos' Theorem may be considered to be a case of Pascal's
Theorem, the two lines of the former theorem being taken as a
degenerate conic.

Had Pascal known of the Principle of Duality he would, of course,
have seen the dual theorem immediately. As it was, the world waited
for over a century before another Frenchman, Brianchon, quite
independently discovered it and today it is known as Brianchon's
Theorem.

Exercise 6i

Discover, and test, with a figure, Brianchon's Theorem. It would be a
good exercise to follow out the dual of the proof.

Exercise 6j

Draw one or more figures illustrating the equivalent for Brianchon's Theorem of the re-entrant hexagons of Pascal's Theorem. (Note. Each of the six tangents of the conic must contain two, and only two of the points of the hexagon.)

Exercise 6l

Prove the converse of Pascal's Theorem; that is, if the three meeting points are collinear then the six points are on a conic.

6.11 Variations of Pascal and Brianchon

There are some interesting variants of the theorem of Pascal according as to whether we let two or more of the points coincide. Let us first consider a pentagon on a conic (Figure 6.14).

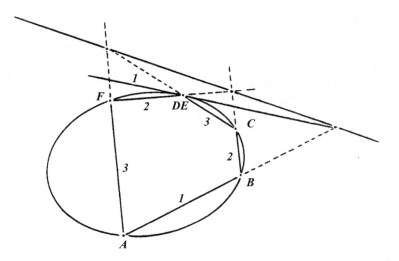

Figure 6.14

We let any one of the points count as a double point — in this case the point $D=E$. Now the sides will be labelled 1 2 3 1 2 3 being, in order, *AB*, *BC*, *CD*, the tangent at $D=E$, *EF*, and *FA*. The tangent counts as one of the sides of the hexagon and we find that Pascal's Theorem still stands.

The relationships become more remarkable when we consider only

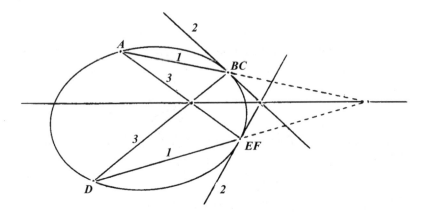

Figure 6.15

four points on a conic. We will consider this to be the hexagon with
sides *AB*, *BC* (the tangent at *B=C*), *CD*, *DE*, *EF* (the tangent at *E=F*)
and *FA* (Figure 6.15).

It is clear that the Pascal line is now on the meeting points of the
two sides *AB* and *DE*, the two diagonals *CD* and *FA*, and the two tangents at *B=C* and *E=F*. But it is equally clear that we could have so
lettered it that the tangents concerned were those at *A* and *D* and we
should have arrived at the same Pascal line, because in doing this we
would not in any way change the meeting points of the sides or of
the diagonals. Therefore, the tangents at *A* and *D* also meet on this
line.

But in quite a similar way we could have so lettered it that the tangents concerned were those at *A* and *B=C*. We should then have
arrived at another Pascal line, which again would pass through the
meeting point of the diagonals and would contain within it the common points of the tangents at *A* and *B=C*, at *D* and *E=F*, and also of
the sides *AD* and *BE*.

We could put this another way. Let us draw the inscribed and circumscribed quadrilaterals at any four points of a conic (Figure 6.16).
We find that the sides of the inscribed quadrilateral meet on the diagonals of the circumscribed one and that the diagonals of the circumscribed one both pass through the common point of the diagonals of
the inscribed one.

It is possible to consider the four points on the conic in another way.
We can letter the points as shown in Figure 6.17.

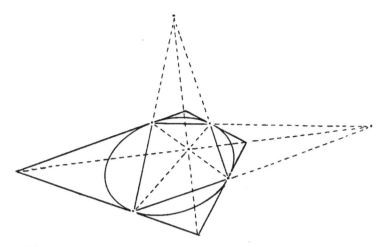

Figure 6.16

The sides of the hexagon would then be numbered 1 2 3 1 2 3 in this order: *AB* (the tangent at *A=B*), *BC*, *CD*, *DE* (the tangent at *D=E*), *EF* and *FA*. We see that the meeting point of the two tangents is collinear with the meeting points of the two pairs of opposite sides. But we could equally well have lettered the same four-point so that the double points came at *C* and *F*. Clearly, we would have then found the meeting point of these two tangents also on the same line, as by

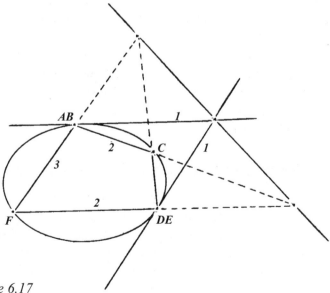

Figure 6.17

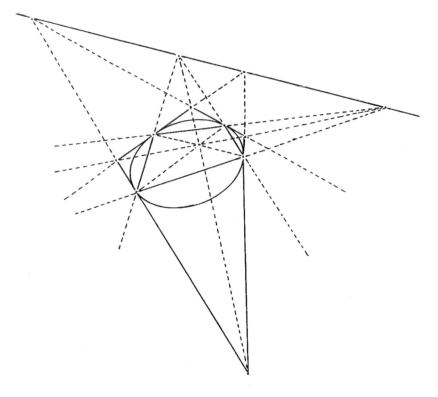

Figure 6.18

so doing we would not have altered the meeting points of the opposite sides.

We can thus add to what we have already discovered about the inscribed and circumscribed quadrilaterals at any four points of a conic, that the four pairs of opposite sides all meet on one line.

The picture, completed as far as this, is important and is worth drawing in full (Figure 6.18).

6.12 The inscribed and circumscribed quadrilateral

There is yet another way in which we could have lettered the four points on the conic, as is seen in Figure 6.19.

We see from this that we have a pair of points, each formed from the meet of a tangent with a side and these points are collinear with the meeting point of the two diagonals. By moving the letters round, while keeping them in the same order we can find three other pairs of points

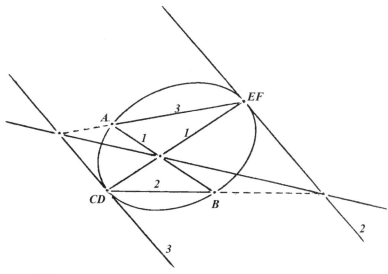

Figure 6.19

formed from the meets of tangents with sides, each of which pair will be collinear with the meet of the diagonals.

By the time we have done this we have drawn all the tangents and, in fact, we have our old picture of the inscribed and circumscribed quadrilaterals at four points on a conic.

The proposition can then be formulated like this. We draw (Figure 6.20) the inscribed quadrilateral at any four points on a conic (heavy) and the circumscribed quadrilateral at those same four points (light). We mark in every point where a heavy side meets a light one; we find that we have eight such points and these points are arranged in four pairs, each of which is collinear with the common meeting point of the four diagonals of the two quadrilaterals. No matter how irregular the original four-point, we now have eight concurrent lines!

We let the inscribed quadrilateral be *ABCD* and the circumscribed one *EFGH*. The meets of heavy and light sides we label 1 2 3 4 5 6 7 8 and find that 1 and 5, 2 and 6, etc. are all collinear with the common meet of the four diagonals which we call point *O*.

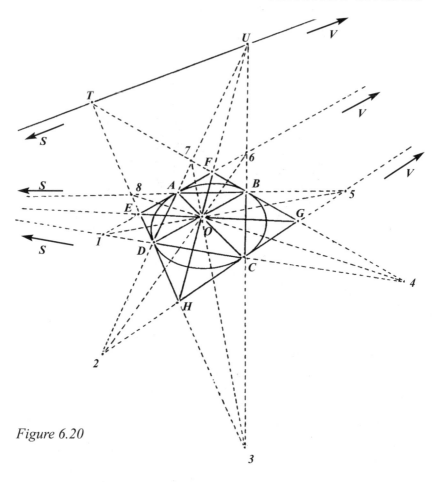

Figure 6.20

6.13 The eight-point conic

We will now prove that, no matter how irregularly the points *ABCD* are placed on the conic, the eight points 1 2 3 4 5 6 7 8 all lie on another conic, which is called the *8-point conic* of the original one.

To do this, we will have to assemble a number of facts. Firstly, note that quadrilateral *ABCD* has its opposite pairs of sides meeting at *U* and *S* and its third pair of 'sides' (diagonals) meeting at *O*. Referring back to Figure 4.6 (page 50), we remember that the four lines *US, U2, UO* and *U3* are harmonic. By projection, it follows that points 3, *O*, 7 and the point where their line meets line *US* are harmonic. Similarly, every pair of numbered points collinear with *O* is harmonic with respect to *O* and the line *US*. This means that these pairs of points transform into one another in a harmonic homology in which point *O*

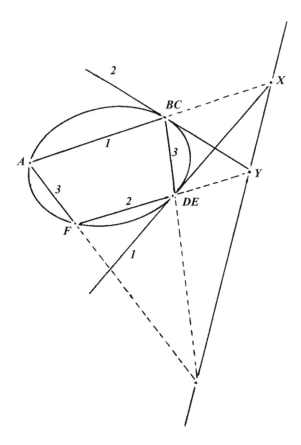

Figure 6.21

is the centre (self-corresponding point) and *US* is the axis, or self-corresponding line (Section 3.3).*

   Thus it appears that opposite points transform into one another in
the homology, 1 into 5, 2 into 6, etc. Therefore, the line 12 transforms
into the line 56 and these lines must meet on line *US*. Thus line 23
meets line 67 on line *US*, etc.

   Another fact which is important for our purpose appears when we
arrange the four points on a conic in yet another way — as in Figure
6.21.

   Notice that the points *X* and *Y* are both formed by the meet of a side
and a tangent and thus are points of the kind which are numbered one
to eight in Figure 6.20. We see now that any two such points which lie
on opposite sides of the heavy (inscribed) quadrilateral must be

* *O* and *US* appear to be pole and polar with respect to the conic; see Chapter 10.

collinear with the meet of the other two sides. Referring to Figure 6.20, therefore we see that 6 and 7 must be collinear with S, 1 and 8 with U, 5 and 4 with U, 2 and 3 with S.

Consider now the hexagon 7 6 3 4 1 8. By considering the facts given above, it will be found that this is a Pascal hexagon, with the line $STUV$ as its Pascal line. Hence all these six points lie on one conic.

Similarly, we will find that the hexagons 1 8 7 5 4 3 and 2 3 4 6 7 8 are also Pascal hexagons with the same Pascal line.

Thus we have three groups of six points, such that each group has all six points lying on a conic. Here are the three groups:

$$1\ 3\ 4\ 6\ 7\ 8$$
$$1\ 3\ 4\ 5\ 7\ 8$$
$$2\ 3\ 4\ 6\ 7\ 8$$

It will be noticed that both the second and the third groups have five points in common with the first; as this is sufficient to determine a conic we see that all eight points 1 2 3 4 5 6 7 8 must lie on one and the same conic.

Exercise 61

Dualize the above. The fundamental figure — an inscribed and a circumscribed quadrilateral at any four points on a conic — will be seen to be a self-dual figure. Any dual qualities which we find will, therefore, apply to the same fundamental figure. Some of the qualities deduced above will be found to be the dual of others, but some of the dual qualities will be new to us. For instance, yet a third conic is 'present,' touching eight lines, which lines can be immediately drawn.

6.14 A triangle on a conic

We have yet another way in which we can make a hexagon on a conic, by having three double points. We can letter any triangle on a conic to represent a hexagon and the sides would then be numbered as in Figure 6.22.

We see that if we have any conic and we put any three points on it, we can draw through these points the inscribed and the circumscribed triangles at these points, and that if we mark the common points of

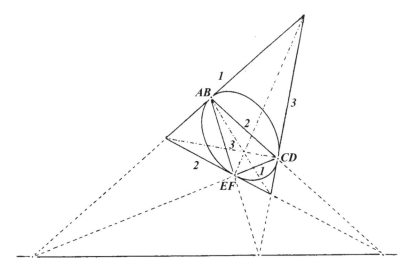

Figure 6.22

opposite pairs of sides in the two triangles we will get three points lying on a line.

Exercise 6m

Dualize the above. The two triangles on the conic obviously form a self-dual figure. The lines of the outer triangle must be considered as double lines and a Brianchon's hexagon is then formed. We find that if we join opposite points of the two triangles we get three lines which are concurrent.*

6.15 The cross-axis

Let us now consider the Theorem of Pappos again (Exercise 2b, page 30 and Section 5.2, page 60). We remember that this reveals itself as a case of Pascal's Theorem on a degenerate conic. We form the figure with the six points ABC and $A'B'C'$ in the usual way (Figure 6.23).

We see the Pappos line, p, formed by the points L, M and N and, in addition, we let the line AA' meet p in K. Now let us take any fourth point of the line x, say, D. We join it to A' and let this line meet p in Q. Join AQ and let it meet x' in D'. Now the cross ratio of the points $ABCD$ equals that of $KLMQ$ and this, in turn, equals that of $A'B'C'D'$. Clearly

---

* Again, this centre and the Pascal line are pole and polar with respect to the conic.

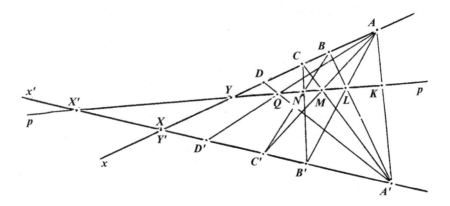

Figure 6.23

then, the projectivity determined by the three pairs *AA' BB' CC'* is also one which includes the pair *DD'* in itself. In other words, every point of line *p*, when projected from *A* and from *A'* gives a pair in the projectivity.

Now let us consider the common point of lines *x* and *x'*. We will call it *X* when it is considered as a point of *x*, and *Y'* when it is considered as a point of *x'*. A moment's thought will show that line *p* will meet *x* in *Y* and *x'* in *X'*. But these two points, *Y* and *X'*, are determined by the projectivity and not by the three pairs *AA' BB' CC'* which we happen to have chosen from it. In other words, we could have chosen any other three pairs from that projectivity and they would have given us the same Pappos line, because this line must, in fact, always pass through *Y* and *X'*.

This line is sometimes called the *cross-axis* of the projectivity. Any pair of the projectivity will project each point of the cross-axis into another pair of that projectivity.

We see now why the theorems of Pappos and Pascal must concern themselves with six points. They concern the determination of a projectivity and for this three pairs of points are always necessary (see Section 5.1). Both these theorems then pick out for us the cross-axis of the projectivity they have determined.

The picture for Pascal's Theorem is very similar, only here, of course, it is a co-basal projectivity on a conic. Clearly, the points where the cross-axis meets the conic are self-corresponding points of the projectivity.

Figure 6.24 shows a projectivity on a conic. By projecting points *BCD* etc. and *B'C'D'* etc. onto *A'* and *A* respectively one finds the cross-axis *p*. *X* and *Y* are the self-corresponding points of the projectivity.

Alternatively one may put *A*, *A'* and *p* down first and by projecting the points of *p* from *A* and *A'* one gets a quick way of constructing a projectivity. Note: if one constructs a projectivity in this way one is still determining it with three pairs of points, for by fixing line *p* one is in effect determining pairs *XX'* and *YY'*; these together with *A* and *A'* are the three pairs of points which one *must* have.

Exercise 6n

Construct such a projectivity. Now it follows from what has gone before that if we project the pairs of the projectivity from any other pair of that projectivity, say *D* and *D'*, the same cross-axis will result. Try it and see.

Exercise 6o

Dualize the above.

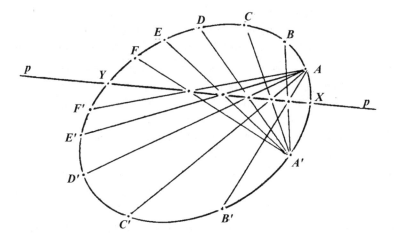

Figure 6.24

6.16 The Steiner circle

These facts can be used in various ways and we give here just one problem as a typical instance.

We are given five points $ABCDE$ and a line x. It is required to construct the points in which the conic which is determined by the five points meets line x and to do so without having to draw the conic.

We choose any two of the five points, say A and B, as raying points. We join AC, AD and AE and let them meet x in C_1, D_1, E_1. Similarly, we let BC, BD and BE meet x in C'_1, D'_1, E'_1. Now from the nature of the conic we know that the pairs $C_1C'_1$ $D_1D'_1$ $E_1E'_1$ are pairs in a co-basal projectivity, in the self corresponding points of which the conic meets line x (Section 6.2). Thus the problem reduces to finding the self-corresponding points of the projectivity which is determined by the three pairs $C_1C'_1$ $D_1D'_1$ $E_1E'_1$.

To do this we put another conic anywhere we like on the plane. For the sake of convenience we generally use a circle. This use of the circle is one which was developed by the celebrated Swiss mathematician Jacob Steiner (1796–1863) and such circles are sometimes referred to as *Steiner circles*.

Now from any point O of the Steiner circle project $C_1 C'_1 D_1 D'_1 E_1 E'_1$ into six points of the circle which we call $C_2 C'_2 D_2 D'_2 E_2 E'_2$ respectively. Join these points in the same way as we did for the Theorem of Pappos (Section 5.2, page 60) and for the Pascal hexagon (Section 6.10). Notice that to do this we join every pair, excepting those which have the same letter in their name.

The three points will be the meets of $C_2D'_2$ and C'_2D_2, of $C_2E'_2$ and C'_2E_2 and of $D_2E'_2$ and D'_2E_2. These three points give us the Pascal line. Let it meet the conic at X and Y. Now X and Y are the self-corresponding points of the co-basal projectivity determined on the conic by the pairs $C_2C'_2$, $D_2D'_2$ and $E_2E'_2$. OX and OY will meet the line x in the self-corresponding points of the co-basal projectivity determined on the line x by the pairs $C_1C'_1$, $D_1D'_1$ and $E_1E'_1$ and these are the points which we wish to find.

Note that this construction only works as long as point O is on the Steiner circle. It is valid by virtue of the facts, that the pencil through any point is projective with the points in which it meets any line (Chapter 4) and that the pencil through any point on a conic is projective with the points in which it meets the conic. We remember, for instance, that the points of a conic have the same cross ratio as the pencil of lines joining them to any other point of the conic. Thus, through point O, we transfer the projective qualities of line x onto the Steiner circle; there, by drawing the cross-axis of the projectivity (that is the Pascal line), we find the self-corresponding points, and, again through point O, we transfer these back onto line x.

To draw such a construction in full is an exceedingly useful exercise. The construction is quite simple if followed step by step, but the finished figure appears very complicated and, as a finished drawing, is not easy to follow. It has, therefore, been thought best not to include a drawing of this in the text, but to leave it to the student to do for himself. In doing this it is a good thing to place the first five points $ABCDE$ on the page in positions which determine a conic which is easily visualized. It is then easy to see whether one's result is a correct one at the end.

7. Growth Measure

7.1 Transforming a line into itself

We transform a line into itself, projectively one-to-one. This means that when we apply the transformation any point A will transform into (or be related with) just one point A'. We will use the letters $ABC...$ exclusively for points before the transformation has been applied and the letters $A'B'C'...$ for those same points after transformation. Each point of the line can then be considered in two distinct ways: as, say, point M, which is a point to which the transformation has yet to be applied, and as, say, point N', which is the result of having transformed some other point N.

We recapitulate the fundamental laws which were described in Section 5.1. We can place any three arbitrary pairs on our line, $AA'BB'$ CC' and say, 'There is one unique transformation which will relate each of these three pairs; and it will, of necessity, contain just two self-corresponding points'.

It follows that if we put two points, X and Y, on our line and say that these are double points (self-corresponding points) of the transformation, there are an infinitude of transformations possible to us. But if we say in addition that any particular point A transforms into another point A' (see Figure 7.1), then the transformation is fixed: the movement of every other point of the line is determined and can be easily constructed.

Given X, Y, A and A' we put any intermediate line through X and another line through Y (Figure 7.2). On the line through Y we mark some convenient point O. Join AO and let it meet the intermediate line in T. Join $A'T$ and let it meet OY in O'.

Now we can quickly find the point B' into which some point B transforms (Figure 7.3). Join BO and let it meet the intermediate line

Figure 7.1

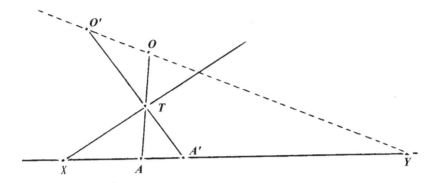

Figure 7.2

in *U*. Join *O'U* and it will meet the base line in *B'*. It is clear that four points *ABCD* would transform into four points *TUVW* which, being projected from *O*, would have the same cross ratio and that these four points, being projected from *O'*, would transfer this same cross ratio to the points *A'B'C'D'*.

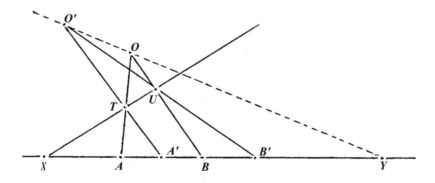

Figure 7.3

7.2 Growth measure

Now we can proceed in this way. We put down our double points X and Y and we transform A into A'. Now we have fixed our transformation. But before the transformation was applied, the point which we are now calling A' was a point which was about to be transformed; let us call it B. And it is clear that the same transformation which turned A into A' would also have turned B into B'. Similarly, the point which, after transformation we are calling B', could have been called C before the transformation had been applied. And it is clear that the same act of transformation which turned A into A' and B into B' would also have turned C into C'. This can be continued ad infinitum (Figure 7.4).

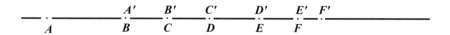

Figure 7.4

We thus get a series of points which is such that one application of a certain transformation will simultaneously move each point one step along the series. By the use of the previous figures of this chapter we can easily construct such a series (Figure 7.5).

Such a *series* of points we call a *growth measure*. There are a few things which we can say about any growth measure immediately.

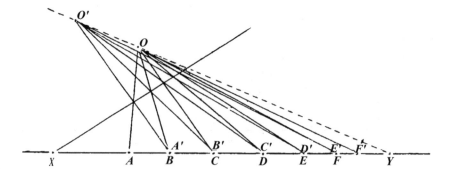

Figure 7.5

1. Every growth measure has two fixed, or double, points and the
    steps of the measure crowd closely together as they approach
    them. It is clear that as we step along the growth measure we
    should never be able to reach point Y; for us it is qualitatively,
    functionally, at an infinite distance.
2. Any two consecutive points, together with the double points,
    have a cross ratio which is constant for the whole growth
    measure. For instance, $XABY$ transforms into $XBCY$ and
    therefore has the same cross ratio and so on with $XCDY$, etc.
3. Also we may note that any four consecutive points have a
    cross ratio equal to that of any other four consecutive points;
    $ABCD$ transforms into $BCDE$, etc. all along the line. This
    cross ratio is not the same as $XABY$ but is related to it in a
    simple way (see Exercise7d).

Exercise 7a

Draw a growth measure, ABC... and then, starting from A and reversing the process construct points in the opposite direction.

Exercise 7b

Like Figure 7.2 draw
— a line l and points X and Y on it
— a second line through X and a third line through Y
— points O and O' on the third line
Now take a point A on l as far to the right as possible. Starting with A,
construct a growth measure. Observe that this growth measure crowds
towards Y again, but from right to left. Reverse the construction, starting with A. Observe that at some step of it, the point jumps to or through
infinity and afterwards the points approach X from left to right.

7.3 The multiplier

Exercise 7c

Draw a growth measure as in Figure 7.5, but erase everything outside line XY. Next draw any line, b, of the plane, conveniently but
not necessarily, through X (Figure 7.6). Draw through Y a line c parallel to line b. Now from any convenient point P on c, project the
points of the growth measure on to line b. It will be seen that we
have ensured that when the growth measure is projected on to line b

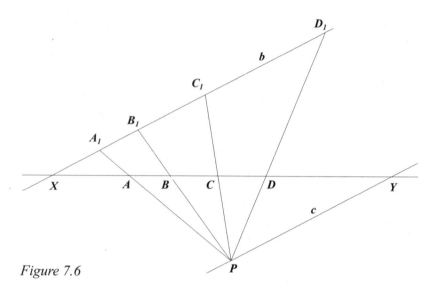

Figure 7.6

we have one of the self-corresponding points (in this case Y) projected to infinity. If line b passes through X, then this double point
projects into itself. Let us name the projections of $ABC...$ on line b
$A_1B_1C_1...$. Notice that the nature of a growth measure is essentially
projective; it is formed simply by repeated projections and its qualities are those of constant cross ratios. It follows that the projection
of a growth measure is always another growth measure. Therefore,
on line b we have a growth measure $A_1B_1C_1...$ whose double points
are X and the point at infinity. Now measure and calculate the ratios

$$\frac{XB_1}{XA_1}, \quad \frac{XC_1}{XB_1}, \quad \frac{XD_1}{XC_1}, \quad \frac{XE_1}{XD_1}.$$

It will be found that these ratios, within the limits of error of our
measurements, are constant all along the line. We find that we have
constructed a geometric series.

Now this is a very remarkable result. We started with a figure which
was in all particulars non-metrical (that is, projective): the original
spacing of $XABY$ was quite arbitrary, as was also the angle of the intermediate line through X and the spacing of O and O'. The growth measure arises without any regard to measurement of any sort at all. The
only metrical concept we introduced was that of the point at infinity

(that is, we made line c parallel to line b); immediately a series of points arises, $A_1B_1C_1$... which is filled with metrical qualities.

We may say that a growth measure is a geometric series seen in perspective, one of the double points representing zero and the other infinity. Notice that these two functions are interchangeable; we could equally well have projected X to infinity and we should then have arrived at a geometric series with the same constant ratio but having Y, or its projection, for zero.

We may say that multiplication, *in its essence*, is not a metrical process at all; it has nothing to do with size or quantity. It is a process which arises very near to the heart of things. We have simply to take the simplest transformation known to us, the projective one-to-one relationship, and apply it again and again in this way. We are then already 'multiplying.' Such a process always has two limits (double points). Take one of these to infinity and the process immediately becomes the ordinary metrical multiplication of everyday life, with the other double point for zero.

In Section 4.1, we said that when we speak of the cross ratio of four points on a line we would take a journey twice from A to C, calling the first time at B and the second at D.

Hence the cross ratio of points $ABCD$, in that order is

$$(ABCD) = \frac{AB}{BC} : \frac{AD}{DC}$$

But we further stated that this is only one of 24 different ways of grouping these four points to give a cross ratio and that these twenty-four ways are grouped into six different values — four cross ratios to each value. For instance, we could travel from D to A, calling first at B and then at C. Our cross ratio of points $DBAC$, in that order, would then be

$$(DBAC) = \frac{DB}{BA} : \frac{DC}{CA}$$

If we now calculate the cross ratio in this way for the double points and any two consecutive points of a growth measure, that is

$$(YAXB) = \frac{YA}{AX} : \frac{YB}{BX}$$

and if we call this the *characteristic number* or *multiplier* of the growth measure, then we will find that this characteristic number is the same as the constant ratio of the geometric series which results from projecting Y to infinity.

In future, whenever we refer to the characteristic number of growth measure we will mean the cross ratio as calculated in this way and it will at the same time be the number by which we are multiplying.

Notice that if we follow the rule for this cross ratio but moving now, not from A towards B, but from B towards A, the cross ratio will read

$$(YBXA) = \frac{YB}{BX} : \frac{YA}{AX}$$

and that this will give a result which is exactly the reciprocal of the previous one. In other words, when we start to move along the series in the opposite direction our multiplier becomes the reciprocal of what it was previously. This accords with what we know of a geometric series: if while going 'up' we are multiplying by, say, five, when we turn and go 'down' we find ourselves multiplying by one fifth.

When we make a growth measure our line becomes, in fact, a picture of our system of numbers (including the 'number' infinity); we establish a one-to-one relationship between the numbers of our number-system and the points of our line. We are free to make either X or Y represent zero and the other double point then necessarily represent infinity. Let us suppose that X is zero and Y is infinity (Figure 7.7). We can now decide whether movement from X to Y in a left-to-right direction (heavy arrow) should represent positive or negative numbers. Having made our decision, movement in the opposite direction will obviously represent the opposite sign. Let us suppose that the left-to-right direction is positive. Then the movement from right-to-left (light arrow) will be negative.

Notice that whereas the functional point at infinity of the line is point Y, the actual point at infinity of the line now represents some finite negative number. Notice also that in a projective number system like this, one has to consider that plus and minus infinity are identical.

$- \twoheadleftarrow X \longrightarrow$　　　　　　　　　　$Y- \twoheadleftarrow$

Figure 7.7

Exercise 7d

Let X and Y be double points of a growth measure on a line l. Let A, B, C and D be consecutive points of it and let $\lambda = (YAXB)$ be its multiplier. Project on a suitable line m such that the projection Y' of Y is at infinity. Now take $X'A'$ as unit, so $X'B' = \lambda$, $X'C' = \lambda^2$ and $X'D' = \lambda^3$. Putting $p = (XABY)$ and $q = (ABCD)$, show that

$$q = \frac{-p^2+p}{3p^2+3p+1}$$

7.4 A negative multiplier

Exercise 7e

Repeat Figure 7.5, excepting that this time the points O and O' are to be put on opposite sides of the intermediate line through X. We stress that this is a most valuable exercise for the student to construct for himself, but because of its importance the beginning of the figure is also included here (Figure 7.8).

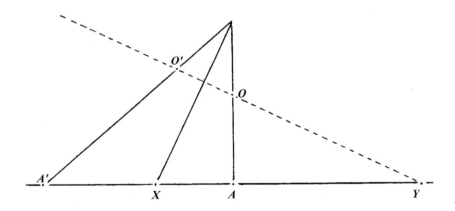

Figure 7.8

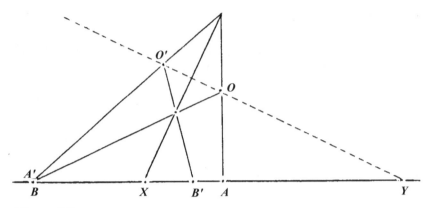

Figure 7.9

We find that *A* projects into *A'* and that the latter is now representing a negative number. Since *A* is a positive number it follows that we must be multiplying by a negative number. Now let us continue the figure. We call the point we have just found, *B,* and we now ask where *B* projects under the same transformation. In other words, we multiply again by the same multiplier (Figure 7.9).

We find that *B'* is a positive number. We have demonstrated geometrically that a minus times a minus is a plus!

Now continue the figure — the remainder of the intermediate line through *X*, below the base line, will need to be drawn. You will find that you have a growth measure which closes upon *X*, first from the positive and then from the negative side, etc. It represents a geometric series with a negative constant ratio,* such as, for instance

$$+16 \ -8 \ +4 \ -2 \ +1 \ -\tfrac{1}{2} \ +\tfrac{1}{4} \ -\tfrac{1}{8} \ ....†$$

---

* For the special case of multiplying with –1 see Section 9.1.

† To each growth measure can be associated a movement of the points on the line (see Section 13.2). If the multiplier is positive each point (except of course X and Y) moves continuously along the line away from X and towards Y. If the multiplier is negative this movement has to be found in the imaginary part of the line. Points move on imaginary loxodromes (spirals) from X to Y. We would like to draw attention to the facts that Rudolf Steiner (1861–1925), without knowing the previous, associated the imaginary with the astral, and pointed out that what are linear movements in the real world become spiral movements in the astral. A most remarkable consistency!

7.5 Speed

Referring to Figure 7.5 again we consider the series consisting of the points *ACE*..., so skipping every second point of our series.

Exercise 7f

Copy the base line with points X, A, C, E and Y of Figure 7.5. Copy also the line through X and the line through Y and only point O on it. Find point O' by the method of Section 7.1. Verify that by this construction C moves to E.

We found a new growth measure. Since one step of the new one corresponds with two steps of the first, we say that the new one has double *speed* with respect to the first. Note that 'speed' is used to compare two growth measures, it has nothing to do with the speed of a moving point. Notice that the multiplier of the new growth measure is the square of that of the first one.

Clearly, moving point O' up results in increasing the speed. More precisely, since the cross ratio ($YAXC$) equals ($YOZO'$) (where Z is the meeting point of the line through X and that through Y), the speed is determined by the mutual position of O, O' and Z on the line through Y. Algebraically the speed is defined as the ratio of the logarithms of the multipliers.

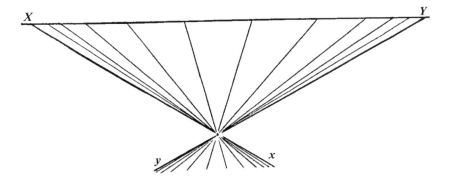

Figure 7.10

7.6 The dual

It is, of course, possible to dualize Figure 7.5, thus obtaining a
growth measure of lines in a point and this is a very good exercise
to do. A similar result, however, can be reached by simply projecting a growth measure of points on a line into any point of the plane
(Figure 7.10).

We see the lines of the measure crowding in towards the double
lines, x and y.

8. Circling and Step Measure

8.1 Circling measure

In constructing a growth measure we imagined a transformation of a line into itself and we produced a range of points *ABCDE...* such that this transformation would turn any, and every, point into its neighbour. In other words, the transformation which turns *ABCD* into *BCDE* is the same one which turns *BCDE* into *CDEF*, etc. Or, in yet other words, any four consecutive points have the same cross ratio as every other set of four consecutive points. And it is clear that any other range of points having this quality must be projectively equivalent to a growth measure. And we have seen that there are necessarily just two points which transform into themselves — the double points of the measure.

Now let us consider two congruent equiangular pencils, with lines, say 10° apart (Figure 8.1).

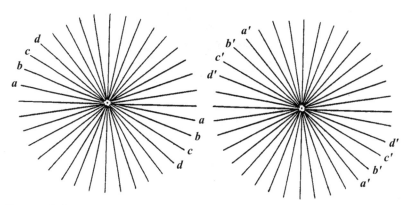

Figure 8.1

We can take any line, *a*, of the left-hand pencil and relate it to any line, *a'*, of the right-hand one. If we want to keep our angles then with line *b*, consecutive to *a*, we have only a little freedom left — we can relate it to either one of the consecutive neighbours of *a'*. After this we have no freedom left — *c* must relate to *c'*, *d* to *d'*, etc. as they are marked in the figure. Now as the pencils are congruent, it is clear that the cross ratio of the four lines *abcd* is equal to that of *a'b'c'd'* and, in fact, any four consecutive lines will have the same cross ratio as their corresponding lines in the other pencil. In other words, we have established a picture of a true projective, one-to-one, relationship between the lines of the two pencils. The line corresponding to any intermediate line of the first pencil, can be found in the second pencil by simply turning through an angle equal to that which the first line has turned. In Chapter 6, we used such projective pencils to construct conic sections of one kind and another.

It is also clear that we could do the same thing with two co-basal pencils; that is, pencils passing through the same point. We could, for instance, decide that each line should transform into the one which is, say, 60° to the clockwise of it (Figure 8.2).

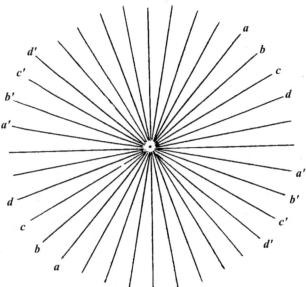

Figure 8.2

Here, to find the line corresponding to any line, we simply count 60° round in a clockwise direction. It is obviously a one-to-one relationship and the cross ratio of lines *abcd* is equal to that of lines *a'b'c'd'*.

Now let us see the same thing, but this time we will make the move from a line to its corresponding line one of 10° (Figure 8.3).

We label any line we like, *a*, and apply the transformation. Line *a* becomes *a'*, 10° round in a clockwise direction (we could equally well have chosen the anti-clockwise direction if we had wished). But this

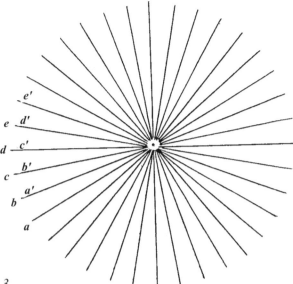

Figure 8.3

line a' was already a line before the transformation; we might have
called it b. And the same transformation which turned a into a' will
have turned b into b'. And so on, in the same way as we proceeded
when we were making a growth measure. It is clear that we have here
a measure of lines which is projectively the same thing as a growth
measure, but we search in vain for the two self-corresponding lines
which we know that a growth measure must have!

Such a measure we call a *circling measure* and we shall find, in due
course, that it has indeed two self-corresponding (double) lines, but that
they are imaginary lines. In fact, we can say that a circling measure is a
growth measure whose double lines have been held back in the realm
of the imaginary. Or we may say that a growth measure is a circling
measure whose double lines have been allowed to precipitate into the
realm of the real, or visible. (We must never fall into the error of believ-
ing that 'real' lines and points are more real than imaginary ones.)

We can project our measure onto a line and so obtain a circling
measure of points along a line (Figure 8.4).

Having produced a circling measure of points on a line in this way,
we could now project these on to any other point of the plane where-
upon we should get a new circling measure of lines in a point, but these
would not, in general, be equiangular at all. Nevertheless, it would be
a true circling measure with all the essential projective qualities of the
original equiangular pencil.

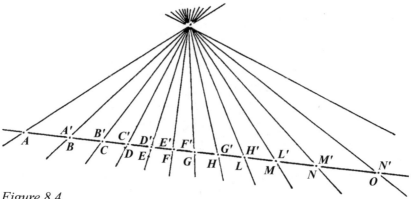

Figure 8.4

8.2 Step measure

There is a limiting case, which stands midway between the growth and circling measures. This is when the two double points (or lines, if we are dealing with a measure of lines in a pencil) merge into one and become co-incident. We can easily construct such a measure by following the method of Figure 7.5 and Figure 8.5. The double point, $X = Y$, is X by reason of the fact that the intermediate line passes through it, and it is Y because we have placed O and O' in line with it. Then we proceed as before.

Exercise 8a

Construct such a measure as this.

You will notice that the points of the measure 'crowd in' towards the double point and space themselves out more and more as they depart from it. If the construction is carried far enough you will find that the

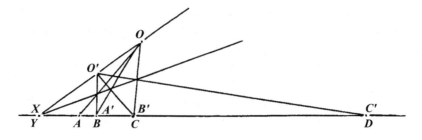

Figure 8.5

measure moves past the point at infinity of the line and, coming back from the 'other side,' starts to crowd in again to the double point. Now draw any other line of the plane, *l*, and through the double point $X = Y$ draw a line parallel to it. Then from any point on this line project the points of the measure onto line *l*. In other words, you will have projected the measure onto line *l* in such a way that the double point $X = Y$ projects to infinity. The surprising fact will then emerge that the points of the measure will project into equal spacing (equal intervals) along line *l*.

We call such a measure, *step measure*. Whenever the double point is at some finite position the step measure is a perspective picture of equally spaced points. As long as we keep our double points real and distinct, we are *multiplying*; as soon as we let them merge we are *adding!* Obviously the former case is the general one; the chances of two points coinciding on a line, unless we do something quite definite to ensure that they will, is exceedingly remote. This is a significant thought in view of the fact that almost all living nature advances by multiplying; nature hardly ever adds! Multiplication is the archetypal process; it is always accompanied by its two distinct guardians, its infinities, between which it runs. Should these double points merge, something like a 'fall' occurs and the process takes on the nature of addition. Should, however, we press through this moment of degeneration to what lies beyond, we find that the two double points, having merged, now separate and pass into the imaginary and the process becomes a circling measure, moving between, or rather (should we say?) under the guardianship of, these two imaginary points.

It is interesting to note that if, in any construction or figure, we change certain elements from real to imaginary, we do not, in general, make any significant change in the qualities of the case. Growth measure and circling measure, however different they may seem in their appearance, are projectively equivalent. But as soon as the two double points become coincident, the whole process moves into a new realm of experience and one which is farther removed from the fundamental things.

Because an ellipse is defined as a conic which cuts the line at infinity in two imaginary points, a hyperbola as one which cuts it in two real and distinct points and a parabola as one which cuts it in two real coincident points (that is, is tangent to that line), the circling, step and growth measures are often called *elliptic*, *parabolic* and *hyperbolic* measures respectively.

Notice that we can put any four points arbitrarily on a line and say, 'These are consecutive points in a measure and, with them, we have uniquely defined this measure.' (Whether the measure is growth, step or circling is not immediately apparent.) For we can put down any three points A, B and C and we can say, A transforms into B, B transforms into C and, because a projectivity is determined by three pairs of related points, I still have freedom to to make C transform into any point I wish; therefore I will make it transform into D. The transformation is now fixed and I have no more freedom left with respect to points E, F etc.

Exercise 8b

Put down four points on a line, with the largest space between the two centre ones (Figure 8.6). This will ensure that they determine a growth measure. Now, by putting down a Steiner circle, projecting the four points on to it and finding the cross-axis of the projectivity on the circle, construct the double points of the projectivity on the line.

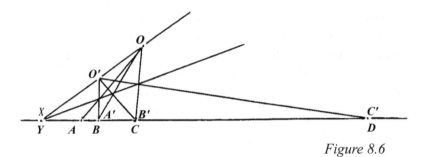

Figure 8.6

Exercise 8c

Repeat the above, with points spaced like in Figure 8.8, having the smallest space in the centre: giving a circling measure. Construct point E of the measure, consecutive to D.

Figure 8.7

Exercise 8d

Put down four equally spaced points (Figure 8.8). Put down a Steiner circle and project the four points onto it. Without further ado, you should now draw the cross-axis of the projectivity. There is only one

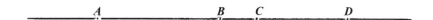

Figure 8.8

place in which it can be and this can be seen immediately. Then, by following the usual construction, as in the two previous exercises, confirm that you have, in fact, put it in the right place. Further confirmation can be had by constructing, with the help of the cross-axis, point *E*.

Do not read this paragraph until you have solved the above problem, or have tried, and failed. Where must the cross-axis go? Firstly, since it is a step measure which we are dealing with, the cross-axis must cut the Steiner circle in two coincident points; that is, it must be tangent to it. Since our four points are equally spaced, their double fixed point must be at infinity along their base line. Hence the answer to our problem: draw a line through point *O* parallel to the base line; where it cuts the Steiner circle in a second point, draw the tangent to the circle; this is the cross-axis we are seeking.

9. Involution

9.1 Breathing involution

When we were considering growth measures we followed the journeying of a point along a line when this line was subjected to repeated applications of a projective one-to-one transformation and we found that, in fact, we were 'multiplying' by some number, the word 'multiply' now being taken in its fundamental projective significance.

There was one obvious special case which we did not mention. It is possible that we transform a point once, into some new position, but that when the same transformation is applied again the point returns exactly to its first position. A moment's thought will serve to show us that in such a case our multiplier must be -1; it is the only possible multiplier which, on a second multiplication, returns a number to its original value.

$$-1 \times +8 = -8$$
$$-1 \times -8 = +8$$

Now let us see what such a transformation looks like from the geometrical point of view. We use the construction of Figure 7.9. As our multiplier is a negative number it is clear that O and O' must be on different sides of the intermediate line through X. If we arrange the positions of O and O' correctly we shall find that a point A transforms into A' which being called B, transforms into B' in such a way that B' coincides with A (Figure 9.1).

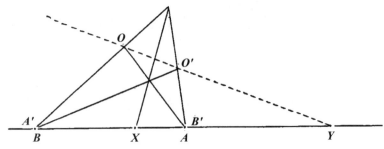

Figure 9.1

Such a transformation is called an *involution*; A and A' are an *involutary pair* or *mates* in involution.

The first thing to notice is that we recognize this figure. It is the harmonic four-point of Figure 4.4. From this we see the fundamental truth of the involution:

any pair of mates is harmonic with the double points.

Next it follows that O and O' are harmonic with Y and the point in which OY meets the intermediate line through X. This, in fact, is the necessary and sufficient condition that puts A and A' in involution. But this condition, when it obtains for one pair, obtains, of course, for the transformation of the whole line. Therefore, we see that if one pair is in involution then every other pair must be also.

This relationship of involution proves to be of fundamental importance throughout projective geometry and it occurs continually in the world around us, it is the closest possible reciprocal relationship. In an ordinary growth measure A is related to B, but B, in the same way, is related to C. B is, in fact, related also to A, but not quite in the same way. A could say, 'B is my relation *after* transformation has taken place' but B would have to say, 'A was my relation *before* transformation took place.' John and Harry are brothers but their relationship is not mutually identical. John says, 'Harry is my elder brother' but Harry has to say, 'John is my younger brother.' The relationship is only identical for each if they are twins. Then they can both say the identical thing of the other, 'He is my twin brother.' Involution is the perfect geometrical example of *twinning*.

You have a perfect example of involution every time you stand before a mirror. You and your image are involutary mates. Supposing the mirror to be silvered on both sides you have only to stand where your image originally was and your new image will be where you yourself originally were. Involution is *harmonic reflection* and is sometimes so-called. If one double point is at infinity, then it becomes ordinary mirror reflection in a plane mirror and the mirror represents the other double point.

Suppose that A and A' are in involution with respect to X and Y as double points. If we now move A towards X we find that A' moves in towards X also. But as A moves away from X, in towards the midpoint of the lesser section between X and Y, so A' moves out towards the midpoint of the greater section between X and Y; that is, to the point at infinity. Because of this reciprocal movement of the pair of mates this kind of involution is sometimes called a *breathing involution*. Because the double points are real it is sometimes called *hyperbolic*.

It is, of course, possible to have an involutary pencil. The easiest way to construct one is to project an involutary range on to any point O (Figure 9.2). Lines a and a', b and b', c and c' are mates. Line x and line y are the double lines, each one mated, or twinned, with itself.

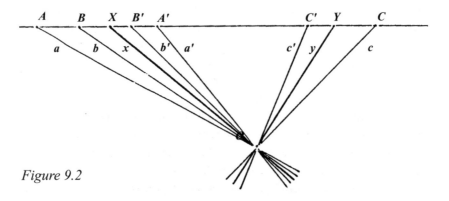

Figure 9.2

9.2 Circling involution

There is an equivalent to this in the realm of circling measure. In Figure 8.2 we produced our circling measure by having our transformation move each line by 60° (had we chosen any different angle, the constant cross ratio of any two consecutive lines with the imaginary double points would have been different). Suppose we chose an angle of 90° movement each time the transformation is applied.

We start with line a and apply the transformation; it becomes line a', 90° onward (Figure 9.3). But before transformation this line could have been called b. After transformation line b, having moved 90° onward, becomes b' and we see clearly that this is identical with the original line a. Again we have an involution! — but here the double lines are imaginary.

This is called a *circling involution*, or an *elliptic* one. We can obtain a circling involution of points by projection on any line in its plane (Figure 9.4).

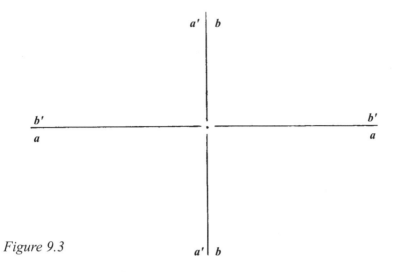

Figure 9.3

Notice that every circling involution of points is symmetrical, having a centre point (in this case *D*) which is the mate of the point at infinity and that if we move one of the points along the line its mate chases it around: hence the term 'circling involution.'

By projecting this circling involution of points on to some other point of the plane we obtain a true circling involution of lines in which the pairs of mates are not, in general, at right angles to one another. To do this would be a useful exercise for the student.

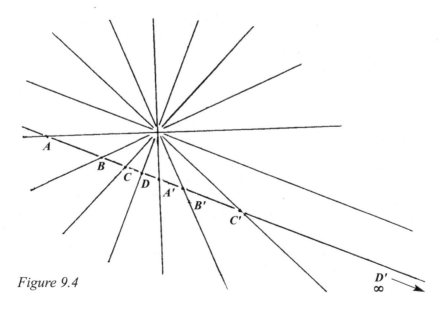

Figure 9.4

9.3 Involution on a conic

Since we can have projective ranges on a conic it follows that we can also have involutions on such a curve and, in fact, the essential qualities of involution are much more easily studied on a conic than on a line.

The construction of an involution on a conic is the simplest matter imaginable (Figure 9.5).

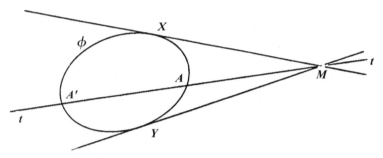

Figure 9.5

Let ϕ be a conic and M be any point of the plane, *not* on the conic. Consider any point A of the conic. A and M determine just *one* line (propositions of incidence, Section 2.2). This line meets the conic in just two points (the conic is a second order curve); one of these is A itself, therefore, it can only meet the conic in *one* other point and it *must* meet it in one other, point A'. As was pointed out with Figure 6.4, this establishes a projective one-to-one relationship between A and A' and as line t turns about M it will sweep out projective ranges A... and A'... But now if we ask ourselves what point corresponds to A' and if we apply exactly the same construction, we find that it brings us back to A. A and A' are a pair in involution. As t turns about M it marks out pairs in involution. The tangents from M to the conic clearly touch it in the double points, X and Y.

This will be found to be the case, no matter how the involution has been constructed. For instance, we might construct an involution of points on a line and then project them onto a Steiner circle, from any point on the circle. Having done so we would necessarily find that the common lines of all pairs of mates would be concurrent at some point. The above figure is the general picture for all involutions on a conic.

If the point M is outside the conic and therefore has real tangents to it, then the involution is a breathing one, with real double points. On the other hand, a circling involution will have its point M (its pole) inside the conic and there will be no real tangents, or double points.

Notice that any pair A and A' are harmonic with respect to X and Y (see Section 9.1) and, in fact, this is the easiest way of constructing four harmonic points on a conic.

Exercise 9a

Let X and Y be two points on a circle and let their tangents meet at M. Through Y draw a line parallel to XM and meeting the circle in T. Let TM meet the circle in S. Let YS meet XM in U. Prove that U is the midpoint of XM.

We can deduce a number of important qualities about involution on a conic.

1) An involution is determined by any two pairs of mates (or by the two double points, which comes to the same thing).

    For the common lines of the two pairs will meet in our point M and with this the rest of the involution can be constructed at will.

2) Any two involutions will have just one, and only one, pair of mates in common.

    For let the poles of the two involutions be M and M'. The line MM' will mark on the conic one pair of mates which will belong to both involutions; there can be no other. Two circling involutions always have a pair of real mates in common and, the same is true for one circling and one breathing. But two breathing involutions may have their common pair real or imaginary. Try it on a small figure and you will see.

There is one more important fact which can be easily seen if we take into account a little elementary Euclidean geometry. Let us consider a breathing involution of lines. We take a circle and put any point M outside it, as the pole of our involution (Figure 9.6).

We let a line through M meet the circle in A and A' and these, of course, are a point-pair in involution. X and Y, where the tangents from

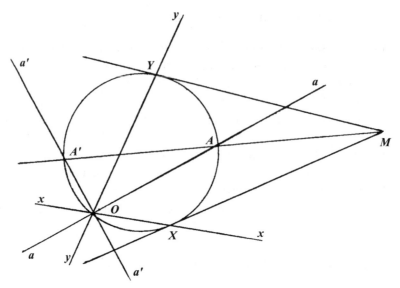

Figure 9.6

M touch the circle are the double points. Now we join these four points to any point *O* of the circle and we get *a* and *a'* as line-pair in involution with *x* and *y* as double lines.

If, as in our figure, the line *MA* passes through the centre of the circle, we know that lines *a* and *a'* must be at right angles (angle in a semi-circle); and clearly this is the only line through *M* which could produce a right-angled pair at *O*.

Also, by Euclidean geometry we know that the arcs *YA* and *AX* are equal; therefore the angles standing on these arcs are equal; in other words, *a* and *a'* bisect the angles between *x* and *y*.

Therefore we may say that in any breathing involution of lines there is always one, and only one, pair of mates at right angles and this pair bisects the angles between the double lines. (Involutions apart, it is true to say that any pair of right-angled lines bisects any pair which is harmonic to it.)

An inspection of the figure will show that matters are the same with circling involutions, excepting for the special case where the pole of the involution lies in the centre of the circle. In such a case, it is clear that all pairs of lines will be right-angled and we come to the circling involution which we constructed in Figures 9.3 and 9.4. Thus a circling involution either has *one* right-angled pair, or *all* the pairs are right-angled.

9.4 To construct an involution

An involution is determined by two pairs of mates. If the pairs do not separate one another the involution is a breathing one.

Exercise 9b

Put down any two pairs which do not separate one another. Project on to a Steiner circle and find the pole of the involution on that circle. Hence find the double points, or any other pair of points on the line. Do the same thing for the same set of points, using a different Steiner circle and confirm that it gives you the same answer.

A circling involution, where the two pairs of points do separate one another, can be constructed in the same way, but there is another and easier way (Figure 9.7).

Draw semi-circles on the two pairs, A and A', B and B'; where these

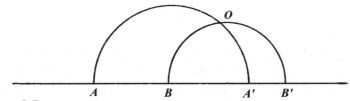

Figure 9.7

meet is the only point in the plane, except for its reflection on the other side of the line, from which both pairs subtend a right angle. This is the centre of a right-angled involution of lines which is in direct perspective with the involution of points. Any other right-angled pair of lines through this point will give a point-pair in the involution on the line.

10. Pole and Polar

10.1 The general case

We have any conic, and a point A, not on the conic (Figure 10.1). Through A we draw any line l, to meet the conic in S and T. Between S and T there will be point A' which is harmonic with A, with respect to S and T. Suppose line l turns about A; S and T will move round the conic; what path will A' follow if it moves so that it always remains harmonic with A with respect to S and T?

Firstly, we notice that as line l moves it sweeps out pairs of points, S and T, which are point-pairs in involution. We call the double points of this involution, X and Y.

We draw any general position of line l (Figure 10.2). We now project the involution on the conic into lines through X. Lines s and t are then a pair of mates in this involution. The double points on the conic, X and Y, project into lines XY and XX which latter, of course, is the tangent at X; that is, XA. But lines t and s being a pair of mates in the involution at X, must be harmonic with respect to the fixed lines of that involution, XY and XA (see Section 9.1). As t, s and XA meet line l in T, S and A, it follows that the point A' which we are seeking, harmonic with A, must lie on XY. Thus the answer to our question is that as line l turns round A, the point A' moves along the line XY.

Point A and line XY are known as *pole* and *polar*. A is the pole of XY, and XY is the polar of A.

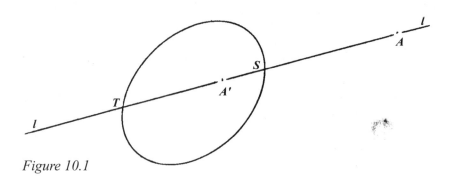

Figure 10.1

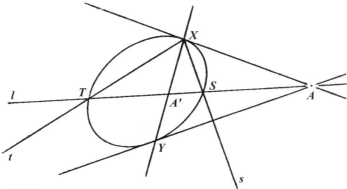

Figure 10.2

We see immediately that in order to find the polar of any point out-side the conic we have just to connect the points at which the tangents from that point touch the conic.

Exercise 10a

If *A* is inside the conic the polar is also defined as the locus of points *A'* that are harmonic with *A* with respect to the meeting points of *AA'* and the conic. Prove that this locus is a line. (Hint: draw three lines through *A* to meet the conic in a hexagon. Find the Pascal line and a harmonic homology.)

If A is on the conic its polar is defined as the tangent at A to the conic

Exercise 10b

What is the polar of the central point of the conic?

Now let *m* be the polar of any point *M* and let *N* be a point on *m* (Figure 10.3).

X and *Y* are the double points of the involution on the conic which has *M* for its pole, and we know that every pair of points on the conic which is harmonic with respect to *X* and *Y* will be mates in this invo-lution, and the line joining them will pass through *M*. Now *N* also determines an involution on the conic, of which *X* and *Y* form a pair of mates, and of which *S* and *T* are double points. It follows then that *S* and *T* are harmonic with respect to *X* and *Y*, and by what has just gone before, the line *ST*, that is the polar of *N*, must pass through *M*.

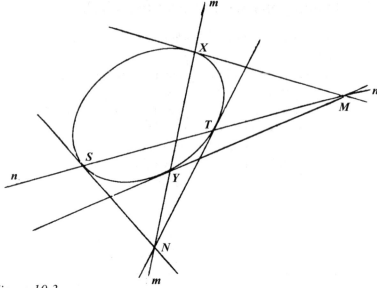

Figure 10.3

Exercise 10c

Verify that this is also true if *M* is on or inside the conic.

It is this which is the fundamental quality of pole and polar; if *N* lies on *m*, then *n* lies on (passes through) *M*. It is a close reciprocal relationship, of involutary quality. We transform *M* into *m*; if we then apply the same transformation to any element of *m* it becomes an element of *M*. In other words, the process of passing from pole to polar preserves incidence.

Notice that if we have three or more points lying in a line, their polars will be three or more lines lying in a point, and this point in which they lie will be the pole of the line containing the original points. Hitherto we have mostly dualized in a free way, making each point and line transform into any line and point we choose, consonant with the projective qualities of the figure. Now we find a more precise, and limited, way of doing it. Each element can transform into its opposite only in conformity with the laws of the polarizing conic. We have, in fact, discovered a way of transforming the plane into itself, one-to-one, in such a way that each point transforms into a line, and each line into a point. Such a transformation is called a *polarity*.

10.2 Pole and polar with respect to a circle

The theorem of pole and polar takes a particularly simple form when applied to a circle.

Draw a circle, centre O, of any convenient radius and put a point A some way outside it (Figure 10.4). Draw tangents from A to the circle, meeting the circle at U and V. Line a is the polar of A and, of course, it is the line UV. Let it meet AO in A'.

We let the line AO meet the circle in S and T. Now we measure OA', OA and the radius OT, which latter we will call r. Within the limits of the errors of our measurement we will find that

$$\frac{OA'}{r} = \frac{r}{OA}$$

That this result is necessary can be seen at a glance when we realize that the two triangles OUA' and OUA are right-angled and similar. From this it is obvious that

$$\frac{OA'}{OU} = \frac{OU}{OA}$$

putting, in each case the smallest over the largest side. And since $OU = r$, this is the formula we arrived at by measuring.

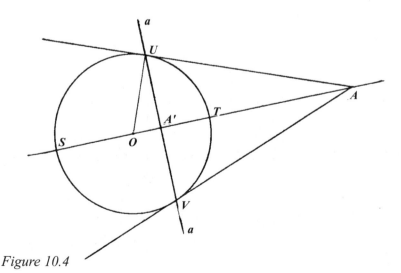

Figure 10.4

If the radius of the circle had been unity this little equation reduces to

$$\frac{OA'}{1} = \frac{1}{OA}$$

In other words, the distance of a point is always the reciprocal of the distance of its polar, the distances always being measured from the centre of the circle in terms of the radius as unity. The circle is sometimes called the *reciprocating circle* and the process of transforming the plane with respect to such a circle is called *reciprocation*.

Notice also that the polar is at right angles to the line *AO*. This is found to be true only in the case of the circle.

10.3 Polarizing a curve

To find the polar of any given curve — perhaps a cubic curve — we put a polarizing conic in some convenient position on the plane. Now we may decide to regard our cubic either pointwise or linewise. If the former, we find the polars of a selection of points along the curve, and these will give us our *polar envelope*. But after this we could, if we wish, draw the tangents of the cubic and find their poles; these will give us the polar pointwise curve, and they will all be found to lie on the polar envelope

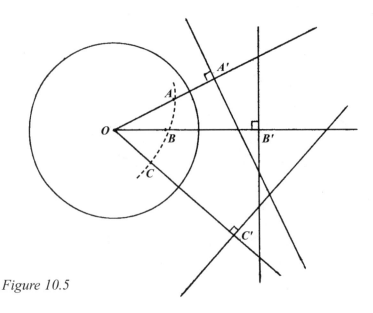

Figure 10.5

which we have just found. The process, however, is cumbersome and the paper quickly becomes covered with construction lines.

Reciprocating with respect to a circle, however, is quicker and easier (Figure 10.5). Let the radius of the circle be unity — say 2.5 cm.

We wish to find the polars of points A, B and C of some curve. Draw the rays OA, OB and OC. Now measure the distance OA, find its reciprocal and measure this as OA'. All these measurements must, of course, be made in terms of the radius of the reciprocating circle as unity. Proceed similarly to find B' and C'. Through A', B' and C' draw the perpendiculars to OA', OB' and OC'. These are the polars which we require. Points A', B' etc. will lie on another, and quite separate, pointwise curve, which is called the *inversion* of the original curve. Plotting comparatively few points will often enable us to draw the inversion quite accurately. To find the polar envelope we then have only to move round this curve with a set-square, keeping one arm of the set-square through O, and the right angle on the curve. In this way we can draw as many tangents to our polar curve as we wish or find the time for, although we may have plotted comparatively few points of the inversion.

Exercise 10d

Draw a 3 cm radius reciprocating circle, and put a small circle α inside it, as in Figure 10.6. Now by drawing rays OAB, OCD etc. find various inverse points for a number of points of label (there will, in

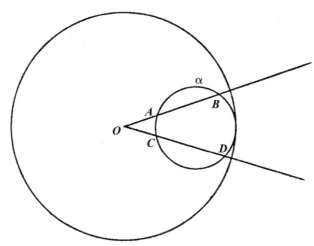

Figure 10.6

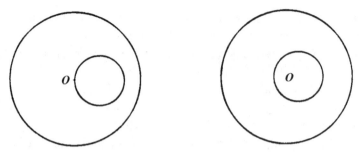

Figure 10.7

general, of course, be two points to every ray). You will find that these inverse points all lie on another circle, which we will call α'. Having found only a few points of α', you can draw it with a compass. Now, moving round with a set-square you can quickly draw the polar envelope of points of α'.

Since order and class are dual qualities, and a conic is both second order and class, it follows that a conic will always polarize into another conic.

Exercise 10e

Repeat the above construction with the circle in the positions of Figure 10.7. It is interesting to note the metamorphosis which both the inverse curve and the polar envelope undergo, as the circle moves gradually to its new positions.

Notice that when the circle passes through the centre of the reciprocating circle it has a tangent line passing through that centre. This means that the polar curve will have a point on the line at infinity. The circle also contains in itself the centre point O, and this means that the line at infinity will be one of the tangents of the polar curve. What curve must the polar envelope be?

11. Dual Properties of Curves

11.1 Behaviour at infinity

Since a conic is a second order curve it follows that it can never cross one of its tangents. For since it meets its tangent in two coincident points at the point of contact, it cannot meet it again without becoming at least third order. Bearing this in mind, if we look at a hyperbola we meet with one of the paradoxes of the infinite. We have seen, Section 6.9, that a hyperbola has two asymptotes, tangent lines which touch it at infinity. It is clear therefore that when the hyperbola returns along its asymptote it must do so along the same 'side' of that asymptote, since it has not crossed it. Nevertheless when we look at the hyperbola, with its asymptotes, it is quite clear to us that it appears to come back along the other side of that asymptote.

The explanation of this apparent paradox is not an easy one for our imagination to grasp. It is a fact that the line, considered in its infinite totality has not got two sides; it is what is called a one-sided entity. And when a curve, passing through infinity, behaves as the hyperbola does (Figure 11.1), it is one which does not cross its asymptote, however much it may appear to have done so. Figure 11.2 represents what the hyperbola is doing to its asymptote at infinity.

But there are curves which treat their asymptotes differently, for instance there is a very common type of curve which behaves something like in Figure 11.3.

Figure 11.1   Figure 11.2

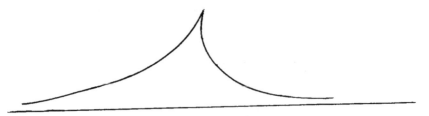

Figure 11.3

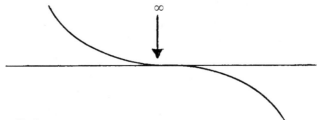

Figure 11.4

This curve runs out to infinity along its asymptote, but returns along what seems to us like the same side. In fact this curve is not only tangent to its asymptote at infinity, but it also crosses over it there. Figure 11.4 indicates what this curve is doing to its asymptote at infinity.

In fact, whenever a curve is related like that of Figure 11.4 to its asymptote — we say of it that it has a *flex* (or *inflexion*) at infinity, and we shall soon see further justification for this.

Figures 11.5 and 11.6 show a useful way to imagine this. We see a

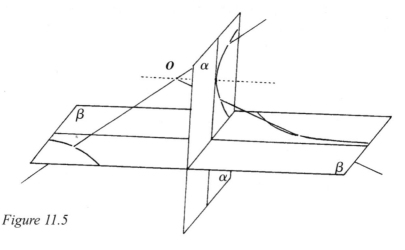

Figure 11.5

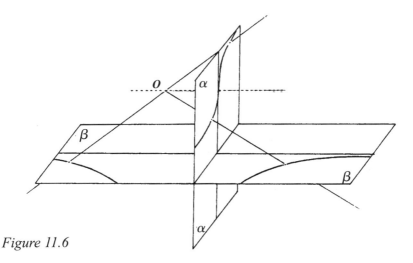

Figure 11.6

horizontal plane β and an upright one α (the *picture plane*). In α we put a curve tangent to a vertical line, and from point O we project the point of contact of this tangent, to infinity along the central line of plane β. In Figure 11.5 we see the projection of the tangent, and in 11.6 the projection of the flex tangent, behaving just in the ways described above.

11.2 Polarizing a set of conchoids

Now let us construct a family of curves, and find their polar curves. We will take a fixed line m and a fixed point M, say, 10 cm from the line (Figure 11.7, where it is done on a smaller scale).

Through M we draw a pencil of lines. Now along each of these lines we measure a fixed distance from m, and mark the point. If the fixed

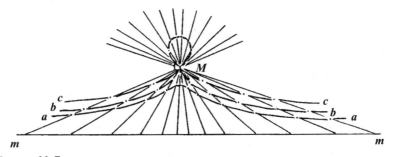

Figure 11.7

distance is less than the distance of *M* from *m*, we get a humped curve (*a* in Figure 11.7). If the fixed distance equals that of *M* from *m*, we get a curve with a cusp at *M* (*b* in figure); and if the distance becomes greater than the distance of *M* from *m*, the curve passes through the cusp into a loop (*c* in figure). These curves are cubics; that is, of the third order. They were studied by the ancient Greeks, and called by them *conchoids*, as they fancied they resembled shells.

When drawing their polars, it is best to draw only one conchoid on each figure. We may start with the humped curve, for instance, and we may put the 2.5 cm polarizing circle anywhere we wish. For a start, it is good to put its centre at point *M*. This will give a symmetrical polar curve. If the polarizing circle is placed asymmetrically, the polar curve will be asymmetrical, but will have all the projective qualities of the symmetrical one. Before starting, notice the nodes, cusps, flexes and bitargents, and any other projective properties of one's figure, and make a prediction of what one will find in the transformed figure. For instance, in the humped curve we clearly have three inflexions: two symmetrically placed on either side of the hump, and one at infinity along line *m*. The new curve will, therefore, be a three-cusped curve. Since the inflexion has its point at infinity along line *m*, its polar cusp will have its tangent along the perpendicular line through *M*; and since this inflexion has its tangent in line *m*, the corresponding cusp will have its apex at the polar point of line *m*. We should deal similarly with the other cusps; the pole of the flex tangent will be the apex of the cusp, and the polar of the point of inflexion will be the tangent of the cusp.

Having determined the situations of the various singularities we may now go on to construct the envelope, using the method of Section 10.3.

It is now most instructive to do the same thing for the cusped curve, on another figure. In the metamorphosis of the original curve, *a*, we see the two inflexions joining up to form a cusp. In our polar curve we shall expect to see two cusps joining to form an inflexion. The original curve *b*, has one cusp and one inflexion. What may we infer about the polar curve? Since with the polarizing circle in the position we have given it, the tangent of the cusp of the original curve *b* passes through the centre of the circle, we must expect to see (or rather not to see!) the two cusps joining at infinity. By doing these two figures with the circle placed in some asymmetrical position we should be able to bring this interesting phenomenon on to the page.

Next we can repeat the whole figure using the looped curve. We now have a curve with one inflexion and one node (crossing point of the loop). We shall get as a polar curve, one with a cusp and a bitangent; the latter will clearly be the line at infinity, since the node is at the centre of the circle.

Now repeat the last construction, but with the polarizing circle moved vertically downwards until it passes through the node. We now get the bitangent on to the page.

Next repeat any or all of the figures any way you wish, putting the polarizing circle in varying positions. Notice that you always get polar curves which are projectively equivalent to those which you have already obtained, but that they *appear* quite diverse and different. By continuing with such exercises one begins to see through the maya of external appearance to the essential projective qualities.

11.3 Polarizing a cardioid

Here is another curve which we can polarize, in the opposite direction — from an envelope to a pointwise curve (Figure 11.8).

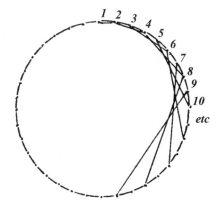

Figure 11.8

We have a circle and divide it into, say, 36 equal parts, which, if we wish, we can number from 1 upwards. Now consider a line which joins points 1 and 2. Let the line move forward so that its righthand end moves two spaces and its lefthand end moves one. In its next position it will join points 2 and 4. In its third position it will join points 3 and 6, etc. We can continue like this until the front end of the line has gone twice round the circle, and the rear end has gone once round. The line will then be found to have enveloped a *cardioid*. Although this curve probably has a different superficial appearance from any of the polar curves which you have constructed from the looped conchoid, a little reflection will show that it is, in fact, projectively equivalent to them all.

This curve can be very easily polarized, with respect to the circle which was used in its construction. Take any tangent to the cardioid and notice the points at which it meets the circle. At these points draw the tangents to the circle, and it is clear that the common point of these two tangents will be the pole of the cardioid tangent. By finding the poles of all the cardioid tangents we can find the corresponding point-wise curve.

12. Pencils and Ranges of Conics

12.1 Conics with four common elements

We know that any two points determine a single line, and it follows from this that if we have just one point there will be an infinitude of lines lying in it; these lines are all related to one another by the fact that they have one point in common. Such a relationship of lines is called a *pencil*. The dual of this is a *range* of points on a common line.

Similarly, any five points, lying in one plane, determine a unique conic. It follows that if we have any four points, always working within the plane now, there is an infinitude of conics which pass through them all. Such a family of conics is called a *pencil of conics*, all the curves being related by the fact that they have four points in common.

It is an exceedingly useful exercise to draw such a pencil, or rather, a selection of the conics belonging to such a pencil. The four points may be arranged in any way one likes, as long as no three of them are collinear. The picture bears a special symmetry if the four points are points of a square.

We refer to Figure 6.9, Section 6.7; this is the easiest way to construct the conic which passes through any five given points.

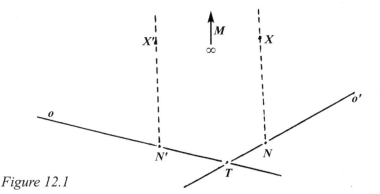

Figure 12.1

We let the four common points of our pencil be X, X', N and N', and for the sake of having a symmetrical figure we will place them on the corners of a square (Figure 12.1). Now the lines NX and $N'X'$ meet at point *M*, so this point will be at infinity, as indicated. Next we take any other point *T*, of the plane, not collinear with any two of the four points already chosen. This point will lie on lines *o* and *o'*. These two lines will, therefore, be *N'T* and *NT* respectively.

Now we can set about constructing our conic according to the method given in Section 6.1.

After this we can take any other point of the plane, not collinear with any two of the four points, and not lying on the conic which we have just drawn. This will give us new lines *o* and *o'*, and, of course, a new conic. In this way we can draw as many of the conics of the family as we wish. If the four points are the corners of a square one of the conics will be the circle which goes through them.

Owing to the limited size of our sheet of paper it is only possible to construct a few points of each conic. However, if the four common points of the system are the corners of a square, all the curves are symmetrical about two known axes, the horizontal and vertical lines through the centre point of the square, and this is a great help in completing the form of each curve.

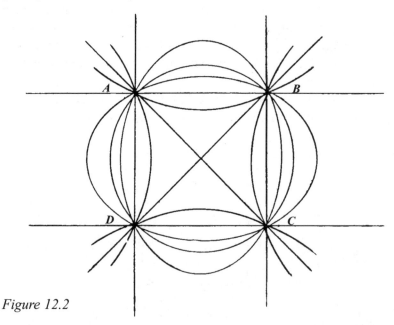

Figure 12.2

Notice that we can move in a continuous and gradual metamorphosis from any one conic through all the conics of the pencil, until we return to the one from which we started. While doing this the conic degenerates three times, once into the pair of lines *AB* and *DC*, once into the pair *AD* and *BC*, and once into the pair *AC* and *BD* (Figure 12.2). These three line-pairs must each be considered as one conic of the system. (In these cases we allow *T* to be collinear with two of the points.)

Exercise 12a

Dualize the above. You will arrive at a *range of conics*; that is, a family of all the conics which touch (are tangent to) four given lines. In order to avoid the page quickly becoming covered with tangent lines it is perhaps best to draw these conics pointwise, but remembering always that the figure is only significant when seen in a linewise way.

12.2 Desargues' Conic Theorem

Another famous theorem attributed to Desargues is the following (his triangle theorem was dealt with in Section 3.1). Let a pencil of conics be cut by any line, *x*. Take any point, *T* of *x*. Together with the four fixed points of the pencil, *ABCD*, *T* determines just one conic. This, being a second order curve, must meet *x* in just one other point *T'* in addition to *T*. We thus have a projective one-to-one relationship in the points of the line, determined by the pencil of conics, in which *T'* corresponds to *T*. Now we must find which point of *x* corresponds to *T'*. It is clear by what has gone before that the conic determined by *T'ABCD* is the same one that was determined by *TABCD*; it therefore follows, that the point corresponding to *T'* is *T*. In other words, *T* and *T'* are a pair in involution on the line *x*. We may now state *Desargues' Conic Theorem*: the conics of any pencil meet any line of their plane in pairs of points in involution. Or, in other words,

> *a pencil of conics determines an involution on any general line of its plane.*

A number of things follow immediately. Clearly the double points of the involution are those where a conic of the system is tangent to the line *x*. Therefore, any line of the plane is tangent to just two conics of the pencil, see Figure 12.3. (Note: we have seen that, in general, one unique conic is determined when five elements, points, lines, etc. are given. Now see that there is an exception if the five elements consist of four points and one line, for in this case there are two conics possible. Dualizing, the same result is obtained when four lines and one point are given.)

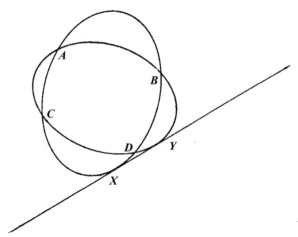

Figure 12.3

12.3 The general four-point and four-line

Next, notice that the line-pairs of the four-point *ABCD* must each be considered as conics of the pencil. Thus *AB* and *CD*, *AD* and *BC*, *AC* and *BD*, each meet line *x* (line *a* in Figure 12.4) in pairs of points in involution, and clearly this must be true for any four-point whatever.

Notice the great generality that has been obtained. If we set ourselves to deal with a square we will find it full of all sorts of special qualities — opposite sides and angles equal, diagonals bisect one another, diagonals equal and at right angles, four different axes of

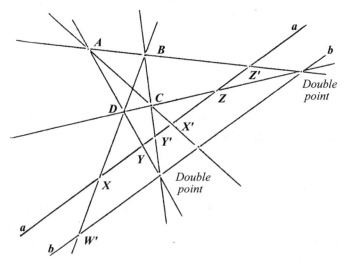

Figure 12.4

symmetry etc. It is hardly surprising that we find all these special facts seeing that we have ourselves imparted to the figure such special qualities when we made it a square. Euclid went further than this and treated chiefly of parallelograms. As a result he imparted less to his figures, and only the first few qualities which we have mentioned for the square, hold for the parallelogram. But suppose we impart *nothing* to our figure, and just have *any* four points in the plane (provided that no three of them are collinear), we might well be excused for wondering whether any special qualities would remain; but if such can be found they are obviously not put there artificially by us, but must be deeply embedded in the very qualities of space itself.

And now we find that any general four-point — of any size or shape whatever — has this wonderful quality, that its three pairs of lines determine an involution on any line in its plane; all the other pairs of mates are shown by the pencil of conics passing through these four points.

If the line in question passes through the meeting points of two of the pairs of sides of the four-point in question (line *b* in Figure 12.4) these points obviously become double points of the involution. The third pair of lines meet in points which are mates in this involution, and these points must therefore be harmonic with the double points. Thus we see that the harmonic qualities of the four-point (Section 4.3) are really only a special aspect of this much more general involutary quality which it enjoys.

Exercise 12b

Examine the dual of Desargues' Conic Theorem and deduce the involutary quality of the four-line.

Several other facts follow from this theorem. Let us take a pencil of conics, and one general point *X* of the plane (see Figure 12.3). We ask ourselves: What are the polars of this point with respect to all the conics of the pencil? We notice first that *X*, together with *ABCD* determines just one conic of the pencil, and we consider the tangent to this conic at *X*. This line we know will be tangent to just one other conic of the pencil, at *Y*; and *X* and *Y* are the double points of the involution which the pencil determines on the line. Now consider any other conic of the pencil; it will meet the line in a pair of points which are harmonic with respect to *X* and *Y*. It follows that the polar of *X* with respect to this conic must pass through *Y*. Here we have our answer: the polars of X with respect to all the conics of the pencil form a pencil of lines, all passing through Y.

Now we can ask another question. We take a pencil of conics, and any general line of the plane. Where are the poles of this line with respect to all the conics of the pencil? First we note that the line will be tangent to two of the conics, at, say, X and Y, and we know that for each of the conics the polar of X passes through Y and the polar of Y through X (X and Y are harmonic with the meeting points of any conic and our line). If we take any conic of the pencil and ask, 'Which line of X will be the polar of Y with respect to this conic?' we could find an answer by joining Y to A and noting where the conic meets this line, say, at Z. Next, we find the harmonic conjugate of Y with respect to Z and A. When we join this point to X we have the polar of Y with respect to that conic. Any other conic of the family must meet YA in a different point from Z (since the five points $ABCDZ$ determine one unique conic) and therefore with respect to this other conic the polar of Y must be some other line of X. It is clear then that we have a projective one-to-one correspondence between the polars of X and the conics of the family, also a one-to-one correspondence between the polars of Y and the same conics of the family. The set of conics thus establishes a one-to-one correspondence between the polars of X and of Y. Clearly, the meets of corresponding lines in these two pencils will be the poles of line XY in the various conics of the family. So we have our answer to this question also: the poles of the line XY in the conics of the pencil will all lie on a conic, which passes through X and Y.

Exercise 12c

Verify this in a figure like 12.3.

13. The Imaginary

13.1 The algebraic background

For the sake of example let us consider this pair of simultaneous equations:

$$y = x^2 - 3$$
$$2x - 3y = 1$$

The first of these, when put on a graph, gives us a conic (it is a second order curve because the highest term is of second power), in this case a parabola, and the second gives us a straight line. Solving these equations, by substituting

$$y = (2x - 1)/3$$

in the first equation we get

$$x = (2 \pm \sqrt{100})/6 = (1 \pm 5)/3$$

so $x = 2$ or $x = 4/3$, whence $y = 1$ or $y = 11/9$.

When we draw the graphs of these two functions (Figure 13.1) we see that they meet at the points which are represented by the answers to the equations, as they have been worked out algebraically.

Let us next consider the equations:

$$y = x^2 - 3$$
$$2x - 3y = 10$$

Solving these in the same manner we get

$$x = (1 \pm \sqrt{-2})/3$$

We are quite entitled to say that there cannot possibly exist a number which is the square root of -2, and, therefore, to pronounce this pair of equations impossible of solution.

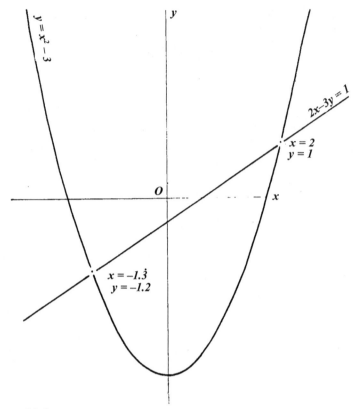

Figure 13.1

As we should expect, when we draw their graphs we find that the line does not, in fact meet the curve in any points (Figure 13.2).

In modern mathematics, however, the possibility has been discovered that one can envisage, and work with, the square roots of negative numbers. Such roots are called *imaginary numbers*, and the basic one with which one works is the number called by the letter *i*, and this is the square root of minus one.

Notice that $\sqrt{(9\times16)} = \sqrt{144} = 12$ and also that $\sqrt{9} \times \sqrt{16} = 3 \times 4 = 12$. Provided that the numbers in the square root sign are *factors*, multiplied together, and that no sign of addition or subtraction appears, then we are entitled to split the square root up like this:

$$\sqrt{(9\times16)} = \sqrt{9} \times \sqrt{16}$$

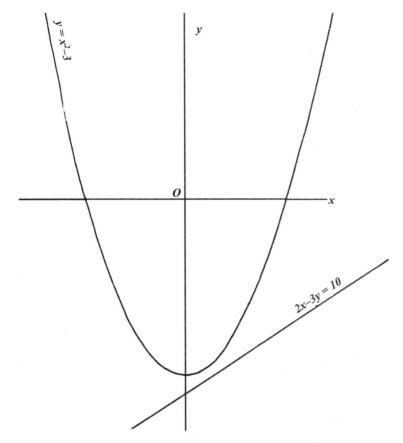

Figure 13.2

Provided no plus or minus enters into the process we get the right answer again.

Bearing this in mind, and remembering that we now have our little

$$
\begin{array}{llll}
x & = & (1\pm & \sqrt{-2} &) \ / \ 3 \\
x & = & (1\pm & \sqrt{(-1\times2)} \ ) & / \ 3 \\
x & = & (1\pm & \sqrt{-1}\times\sqrt{2} \ ) & / \ 3 \\
x & = & (1\pm & i\sqrt{-2} &) \ / \ 3 \\
\end{array}
$$

symbol $i = \sqrt{(-1)}$ to work with, we can get a pair of answers even to our second pair of equations.

A number such as $1 + i\sqrt{2}$, consisting of a *real part* (1) and an *imaginary part* $(i\sqrt{2})$, is called a *complex number*. Our traditional numbers, by contrast, are called *real numbers*, and they form a part of the complex numbers, viz.

$$-3 = -3 + 0i$$

The non-real numbers are called *imaginary*, their coefficient of i is non-zero. Numbers without a real part, like $2i$ ($= 0+2i$), are called *pure imaginary*.

Once we have found an answer for the above pair of equations we have to look for the pair of points in which their graphs meet! We have to say that the line meets the curve in two *imaginary points*. The perception as to just what these imaginary points are, we owe to the famous geometer Von Staudt, and it is his theory of the imaginary which we are now going to explore.

Seeing that the idea of imaginary points and lines finds its genesis in the algebraic realm, and that the algebra gives us full power to work accurately and concisely with them, it might be asked why we should need to carry our studies further, in the way of this book. To answer this we can surely point out that it is always good to come to grips with the inner nature of that with which we work. Even the most abstract algebraist feels the need to have some spatial representation of his imaginary elements. To do this he establishes a symbolic representation, in which each real point of his plane stands for one complex number; a plane used in this way he calls an *Argand*, or a *Gaussian plane*, and its use can be very helpful for understanding, and computational, purposes. But we must be clear that this representation of imaginary elements by real points is purely symbolic and abstract. It is useful for handling them, but it gets us in no way nearer to an appreciation of what they *are*. It is this latter task on which we are now going to embark.

13.2 The group of growth measures

In order to work our way to a perception of how this parabola, or indeed any other kind of curve, can have a part of its being on a line which does not meet it in any visible points we shall have to refer back to some of the things we learnt in Chapters 7 and 8. First let us remember the growth measure, which is pictured in Figure 7.5 (page 100). We have a steady movement of a point, from A to B, to C, etc. and this movement takes place between two fixed or invariant points, X and Y. This movement is one without end: X and Y are infinitudes; it has taken an infinite number of steps to emerge from X, and it cannot reach Y in less than another infinite steps. We found that the nature of this movement is one which has within it the quality of the process of multiplication. If we project either one of the fixed points to infinity, we find that all the others take up positions along a true geometric series, with a quite definite multiplier, which we in fact quite unconsciously choose when we put down the determining elements of our construction the intermediate line and the points O and O'. This multiplier is a characteristic number of that particular growth measure, even when both invariant points are on the page, and the spacing of the points of the measure does not show as multiplication in the form in which we ordinarily experience it; in such a case this multiplier shows itself as the cross ratio of any two consecutive points of the measure when taken together with the two invariant points; and this cross ratio is constant all along the measure.

Now if we look back again to Figure 7.5 we see that we could have started to make our growth measure starting from some other point of the line — say very close to point A; and we should then have produced another growth measure moving between the same pair of fixed points, and with the same multiplier; in fact since there are an infinite number of points between A and B, we have the possibility of an infinite family of growth measures, all moving between X and Y and all with the same multiplier.

From now on — when we speak of a growth measure we shall mean not the individual points which are pictured in our figure, but the whole growth-measure-movement along the line — which expresses itself in the totality of growth measures which all have that same multiplier, and which move between the same invariant points. The entity, growth measure, will be one simply of *movement*, and the points which we construct on our page will be no more than a rather crude and imperfect way of picturing to ourselves the qualities of this movement. By playing with the construction of Figure 7.5 we can easily convince ourselves that this movement-organism, which we are calling 'growth measure,' occupies the whole length of the line; it goes on not only in the interior sector, between X and Y, but also similarly in the exterior sector.

Speaking in a very rough and ready manner we can say that the spacing of the points in our measure gives us a sort of hint of the 'speed' of the movement in that neighbourhood, where the word 'speed' is not employed in the continuous infinitesimal sense of normal use. Speaking in this sense we can say that the speed of the measure is greatest around the midpoint between X and Y, and slows down to zero at the invariant points themselves.

But now let us go back once again to Figure 7.5. Suppose we repeat the whole thing, just as it stands, except that we move the point O' a very small distance along the line OY towards O. We shall then get a new growth measure (meaning a whole growth-measure-movement) still between the same invariant points X and Y, but now with a slightly smaller multiplier. Since there are an infinite number of positions along the line OY which the point O' can occupy it follows that we have the possibility of an infinite set of growth measures, all moving between the points X and Y, but each with a different multiplier.

When we come to study the theory of sets we soon meet a type of set the members of which are related to one another in a very definite way. And when we find a set, the members of which are related to one another in this particular way — when they are subject, as it were, to this very special social organization — then the set is called a *group*. And in modern mathematics, groups such as these, considered as whole things, have proved to be entities in their own right; and they have shown themselves to be highly significant entities, not only in pure mathematics but in the physical sciences also.

There is not space here to go into the intricacies of group theory, where we would be 'doing our sums,' not with numbers, or letters, but with whole groups of entities, each group acting as an entity in itself. And for our present purposes there is no need. Suffice it to say that the members of this infinite set of growth measures are interrelated one with another in just this sort of way. All the transformations of the line having X and Y as their invariant points form a cohesive, intricately interwoven social organism of entities — a group of growth measures — and this group, taken as a whole, is an entity in itself; in fact it is the fundamental entity with which we are going to be concerned.

Just see how far we have come along the road towards pure sense-free thinking! The original elements of our geometry — point, line and plane — are surely conceptual in nature; such things have never existed, nor could they ever exist, in material fact. Yet our ideas of them are taken, in the first place at any rate, from our sense perceptual experience. We are quite clear in our minds that we can never see an actual point on our figure, but at least we can see (or we think we can see) the place where we *would* see it if it were big enough to be seen! Such elements we might describe as being quasi-sense-perceptible. But when we come to consider growth measure we are confronted by a whole set of such quasi-sense-perceptible elements, and because it is an infinitely large set we cannot try even to represent more than a tiny part of it on our page; and when we have penetrated through the illusion of these clumsy dots which we make with our pencil to the real concept of the infinitude of points which they represent, we still have not reached the final reality; for these points, by their spacing, are only representing to us something more ultimate still — a whole system of movement, of transformation. And lastly we come to the group of growth measures, a compact highly organized family of an infinite number of such systems of movement which expresses the intrinsic qualities of the process of multiplication by all the non-zero numbers there are.

When we consider this vastly intricate organism of movement we see that its component parts are systems of growth measure movement; and if we ask what it is that binds these various parts into one coherent whole we can say, in a very descriptive manner, that it is the fact that all these movement systems, the fast ones with high multipliers, and the slow ones with multipliers near to unity, share the common quality that they slow down in the same two neighbourhoods, and finally come to rest in the same two invariant points (double points). Here, in the fixed points, we see the inert, crystallizing out of that which is for ever in movement, the manifest being born out of the unmanifest, the (at any rate quasi-) visible coming to light out of that which, by its very nature, must remain forever invisible.

13.3 The group of circling measures

If we now look back to Figure 8.4 (page 112) we shall find a very similar state of affairs. In this circling measure of points along the line we have a set of points which manifest exactly the same fundamental projective quality as the growth measure — that is, the cross ratio of any four consecutive points is equal to that of all other sets of four consecutive points; it is invariant all the way along the line. But in some more superficial qualities there appear to be differences. For instance, if the equi-angular pencil of lines which produces the circling measure, turns just 10° per step, then after exactly

18 steps of the transformation point A returns to the original position which it held. This is the kind of thing which no growth measure can do, except just for the involution. Such a circling measure is called an 18-*cycle*. Clearly the line holds the possibility for an infinite family of 18-cycles, since we could have started our circling measure at any one of the infinite number of points which lie between A and B. The totality of all these circling measures we can describe as a circling-measure-movement along the line; this system of 18-cycled circling measures is the exact equivalent of a growth-measure-movement, all the measures of which have the same multiplier. But by varying the angle turned at the centre of our projecting pencil we could have had 19- or 20-cycles or 100-cycles, or indeed circling measures associated with any number from 2 upwards.

In addition to this we could turn our line by any other angle we wish, per step of the measure. Most of these angles could never bring the point back to the position it originally held; they would be irrational to the total 180° which are to be turned, and the point would continue to go round and round the line, forever, always occupying a new position on the line. Such a measure could not be called 'cyclic' in a true sense, but it would nevertheless be a true circling-measure-movement, of essentially the same nature as the so-called cyclic measures.

And the totality of all these possible circling-measure-movements forms a *group of circling measures*. It is a close-knit entity in exactly the same way as a group of growth measures. In their most fundamental projective qualities these two concepts are equivalent, indeed almost identical. But in their more superficial appearance they have one big difference: whereas the growth measures of the group all come to rest in two invariant points, the circling measures do not. We see them slowing down in some middling neighbourhood of the measure and then, before they can come to rest, they start to quicken up again. It is as though they try to manifest their fixed points, but just fail to achieve it; it is like a being coming down towards the earth but which, before it actually touches the ground, hovers for a moment, and then retreats. We can get the feeling that this group of circling measures holds the same possibilities within it as does the group of growth measures, but the invariant points never 'freeze out' into actuality; they are held back in the bosom of the movement; they are the *imaginary fixed points* of the transformation, forever invisible.

But we must come to much more exact concepts than just having a 'feeling' about it. We must next ask ourselves what it is that binds together all the measures of a circling group. To answer this we look again at Figure 8.4. A perpendicular dropped from the centre of the pencil meets the base line at point *F*; this is the point where the movement of the circling measure along the line becomes slowest; immediately on either side of it occur the smallest intervals of the measure. This point is called the *centre* of the circling measure. If we take the two rays from the centre of the pencil, which make angles of 45° with this perpendicular, they will mark off for us that section of the base line which holds just half of the points of the measure. The width of this section is called the *amplitude* of the circling measure. In Figure 13.3, *C* is the centre of the measure and *MM'* shows us the amplitude. The centre shows us the *position* of the circling measure whilst the amplitude shows us the *quality* of the movement; a wide amplitude tells us that the movement slows down very little, but a narrow amplitude shows us that the imaginary fixed points which are held in the movement are very near to being born into actuality. This narrow amplitude would come of course when the centre of the projecting pencil, *O*, is very close to the base line of the circling measure. If the amplitude becomes zero, then a pair of co-incident fixed points appear at *C* and the circling measure becomes step measure. The fixed points have then become real.

The condition that our set of circling measures should be bound together in a group is that they should all have the same centre and

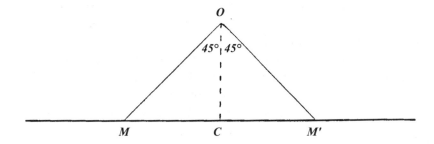

Figure 13.3

amplitude. The fact that there should be two determining conditions for this might have been guessed from the work which we did in Section 13.1. The real part of the solution which we found for the second pair of equations is represented here by the position of the centre, and the imaginary part of the solution by the amplitude, which shows us the quality of the circling movement.

Another thing to notice about the solution which we there achieved is that not only is it double in nature (having a real and an imaginary part) but that there are in fact two solutions: $(1 + i\sqrt{2})/3$ and $(1 - i\sqrt{2})/3$. The fact that two solutions should appear is quite general in such cases and is due to the fact that square roots always appear in pairs, whether they are real or imaginary; for instance, if $x^2 = 9$ then $x=3$ or $x=-3$.

And now we are ready to ask the essential question: what exactly are these imaginary points? To answer this Von Staudt draws our attention to the fact that we are looking for the invariant elements in the group of circling measures. How can we possibly find an invariant element where *everything* is moving? Well, the one thing, and the only thing, which is unchanging throughout the whole process is the quality of the movement itself. We think of the myriads of circling measures concerned, of the whole close-knit group of circling movements, and we realize that all this goes on unchanged while every single point of the line is moved. But this whole movement organism can be conceived in two distinct ways, as moving from right to left, or from left to right — if the first is the $+i\sqrt{2}$ then the second is the $-i\sqrt{2}$ (or which way around we wish). We need to be very careful when we use the verb 'to be'; whether it is better to say that an imaginary point is the group of movements, or something even more ultimate yet but which is carried as it were in the bosom of the movements, it is hard to say. But in either case I am sure that we come very close to the essential reality.

So, the imaginary points are, or are carried by, richly complicated groups of circling movements. Does this sound too far fetched for belief? And how does this bring us any closer to understanding how a curve can have anything in common with a line which lies wholly outside itself?

Well, the proof of the pudding is in the eating. In our following studies we shall see, again and again, how these groups of movements are associated together, how they can be related one with another, in just the same ways as ordinary points and lines can be, and how, in every case the results are exactly analogous — circumstantial evidence perhaps but in the long run this can pile up to make a very convincing case. And in the course of this we shall discover how very much our concepts of what constitutes a point, a line, a curve, etc. need to be widened and enriched.

Exercise 13a

Try to dualize the contents of this section. Define the concept of an imaginary line. An imaginary point is said to be on an imaginary line if their movements are perspective (see Section 16.2). Reflect on incidence. What are meet and join of (imaginary) points and lines? See also Section 17.3.

14. The Nature of a Curve

14.1 The skeleton of a circle

In this part we shall work with circles because they are easy to draw, but everything which we will do would apply exactly, and in detail, to any of the conic sections — ellipses, parabola, or hyperbolas. Other higher order curves would obey the same general principles, but the details would be more complicated.

So we will start by doing an exercise with the circle. We will describe it in detail and illustrate it (Figure 14.1), but the reader is strongly advised to make his own drawing step by step.

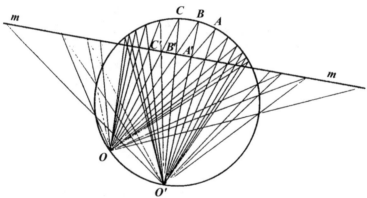

Figure 14.1

We cut our curve by a line *m*, and on the curve we place two points, *O* and *O'* anywhere we like, but not on *m*. And now, starting with a third point, *A*, of the curve, we proceed like this. Draw *OA* and let it meet line *m* at *A'*. Draw *O'A'* and let it meet the curve at *B*. Draw *OB* and let it meet *m* at *B'*. Draw *O'B'*, and let it meet the curve at *C*, etc. Continue like this indefinitely, or until you get tired. For a change you can start again from *A*, but join first to *O'* and then to *O*, etc. This will take you backwards along the series. The exterior portion of the line *m* can be treated similarly. When you have done this you will find that your circle has engendered on line *m* a growth measure (and not only this, but there appears a similar growth measure, with the same multiplier, on the circle itself). Now if we keep *O* fixed, and move *O'* just a little way round the circle we shall find that a new growth measure appears on line *m*; it has the same fixed points as the original one, but a different multiplier. Seeing that there are an infinity of different positions which *O'* can occupy on the circle it is clear that the circle is engendering on line *m* an infinite set of growth measures. Now this set turns out to be a group of growth measures of exactly the kind which we described in Section 13.2. It is bound together by the fact that all the measures have common fixed points, and these are the points which are common to the line and circle. And here is the first place where we are going to have to enlarge our concepts. We have only to cut this circle by any line whatever, and we see, spawned upon it as it were, such a group of growth measures. And we are going to say that these growth measures are truly *part of the circle*. And now consider — there are infinitudes of lines which can cross this circle, and on each the circle possesses an infinite group of growth measures. The circle becomes for our imagination an activity, a close-knit, highly complicated and incredibly rich organism of movement. But all these groups of movement are purely conceptual in being, by their very nature for ever beyond the bounds of sight; they only become quasi-sense-perceptible in their fixed points, where the movement has ceased. The circle truly is activity, and as such is invisible; it comes to sight where the movement freezes; what we can 'see' is only the skeleton of the true circle.

Exercise 14a

Repeat these constructions on an ellipse, a parabola and a hyperbola.

14.2 An organism of movement

Next we will do the same thing again, but this time having our line outside the circle; that is, having no real points in common with it (Figure 14.2).

We find that the circle is here engendering on the line a whole group of circling measures, and if we regard this group of movements as carrying a pair of imaginary points, then we have found the pair of imaginaries which this circle shares with this line.

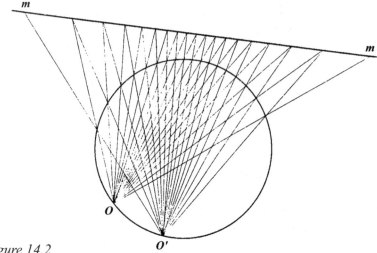

Figure 14.2

Exercise 14b

Try the above on the other types of conic section.

Exercise 14c

Do the whole thing again making line *m* tangent to the circle. You will find that you are constructing a step measure with its two co-incident fixed points (Section 8.2). But in Section 2.8, we learnt to see a tangent as a line which has two co-incident points in common with the curve; the whole thing hangs together.

So we see the circle as a vast richly-woven organism of movement, spread out over all the plane, endowing every line with an infinitude of growth or circling measures, but coming to visibility only in that little ring of points where its movement has died.

And a similar situation holds for every conic section.

14.3 The meet of two circles

We know that two circles can meet in two points, and then they have a common chord (Figure 14.3). But what if they are situated as in Figure 14.4; is there a common chord then?

In order to investigate this possibility we will recall a little-known theorem of the old fashioned Euclidean geometry. We will not attempt

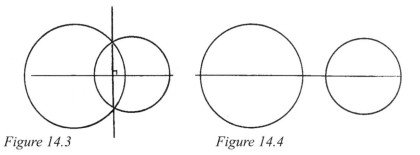

Figure 14.3   *Figure 14.4*

to prove it, but it would be a good thing for the reader to test it in one or two drawings — and convince himself of its truth. Suppose we have three intersecting circles (Figure 14.5) and draw their three common chords; the theorem states that we shall always find these three chords to be concurrent, no matter how variously the circles are disposed.

Another fact we shall need for this exercise is the well-known one that the common chord of two circles is always at right angles to the line of centres (illustrated in Figure 14.3). We shall print most of the resulting drawing, but the reader is strongly urged to make his own as he reads through the paragraph that follows.

First we put down our two circles, making sure that they do not meet in real points, and having one of them distinctly larger than the other for the sake of extra interest. Now we wish to find what line of the

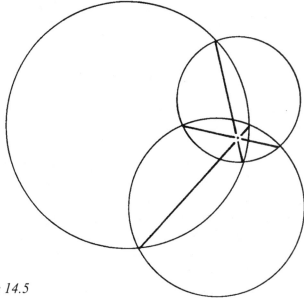

Figure 14.5

plane could possibly serve as a common chord — if any. And we shall
work by analogy with the real case illustrated above. We cut our circles
by any third circle, and we can then immediately draw in two common
chords. The common chord we are looking for ought to be concurrent
with them; that is, it must pass through their common point. It also
ought to be perpendicular to their line of centres. Therefore there is
only one place we can put it — as shown in our Figure 14.6 (line *c*).

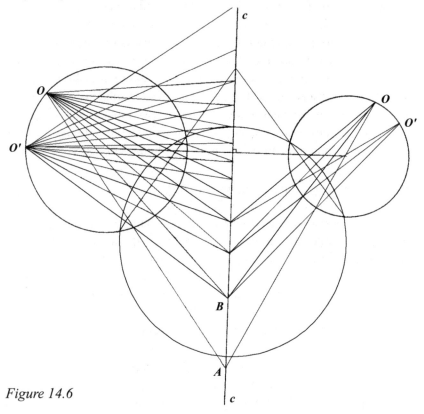

Figure 14.6

If these two circles have a common chord it ought to be this line. Now
we must test to see whether in fact these circles find themselves with
common imaginary points along this line. We put two points *O* and *O'*, in
any convenient position on the larger circle, and working from them we
construct a circling measure on line *c*. This will be just one of the infinite
group of circling measures which this circle possesses on this line. We
next wish to test and see whether the smaller circle also possesses an
identical circling measure along the line. To do this we must choose an

appropriate pair of points on the smaller circle for O and O'. This is done in the following way. From the first point of the circling measure which we already have (point A) draw a line to meet the smaller circle in two convenient points, and one of these we call O. From the next point, B, of the circling measure, we draw a line through the point where OA meets the smaller circle, and this gives us our point O'. Now from these two points we continue to construct the circling measure which the smaller circle makes on line c. If it is done with care and accuracy (and no cheating!) we shall find that the new circling measure is identical with the original one all along the length of line c. It is instructive to try the same exercise along various other lines of the plane; in every case it will be found that the circling measures diverge more and more wildly from one another as one proceeds with the construction. On this line, and this line only, of all the finite* lines of the plane, these two circles meet. Here their movement organisms merge and unite; here they have community; line c is truly their common chord, bearing their common imaginary points.

14.4 The meet of two ellipses

To find the meet of two ellipses is a little harder, for they meet one another in four points, and have no less than six common chords (Figure 14.7). And neither the Euclidian theorem about the intersection of three circles, nor the right-angled properties of the circle still hold.

But if we can draw any real common tangents to the two curves we can solve the problem fairly easily. Let us think back to the transformation of homology which we worked with in Section 3.3. This transformation turns one conic into another, and since the axis of the homology, line l, is a line of self-corresponding points, if this line meets the first conic, it clearly must meet the new one in the same pair of points; it will be their common chord. So if, having two conics which do not meet in real points, we can put them into homology with one another, it would be interesting to see whether the line l of this homology is in fact acting as their common chord; that is, whether it contains a pair of imaginary points which are common to both curves. Let us try.

First we draw two common tangents and let them meet in point O of our homology (Figure 14.8). If the tangents touch the conics in X, Y, X' and Y', then X transforms into X' and Y into Y'. Also O, being self-corresponding, transforms into itself. Now later, in Section

---

* We will see (Section 15.5) that two circles also share the line at infinity.

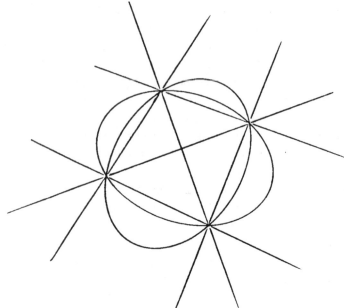

Figure 14.7

21.1, we shall show that any collineation (of which homology is a special case) needs four pairs of points to determine it. We need one more pair. Draw any line through *O* to meet the conics in *A* and *A'*. We let *A* transform into *A'* and our homology is determined. *AX* and *A'X'* will meet on the line *l*, and so will *AY* and *A'Y'*. Join these two points and we have the line *l* which we are seeking. We still have to ask ourselves whether, with this homology, the whole of the first conic will transform into the whole of the second. A moments thought will assure us that it must. For consider the pencils of lines in *X* and *Y*. The first conic induces a projectivity in the lines of these pencils and this projectivity entirely determines the whole form of the conic (Section 6.1). This projectivity is completely determined by any three pairs of corresponding members (Section 5.1). Three such corresponding pairs can be immediately seen:

— line *XX* (that is, the tangent at *X*, viz. *XO*) corresponds to *YX*
— *XA* corresponds to *YA*
— line *XY* corresponds to *YY* = *YO* (that is, the tangent at *Y*)

For the second conic we should have to write:

— line $X'X'$ ($=X'O$) corresponds to $Y'X'$
— $X'A'$ corresponds to $Y'A'$
— line $X'Y'$ corresponds to $Y'Y' = Y'O$

Examining this we see that in every detail the homology transforms the correspondences of the first set into those of the second. This is sufficient to tell us that this homology will transform the whole of the first conic into the whole of the second. Now we have found our line l. It simply remains to test whether the two ellipses possess identical imaginary points along this line. If the reader tries this, using the same method which we used for the common chord of the two circles, he

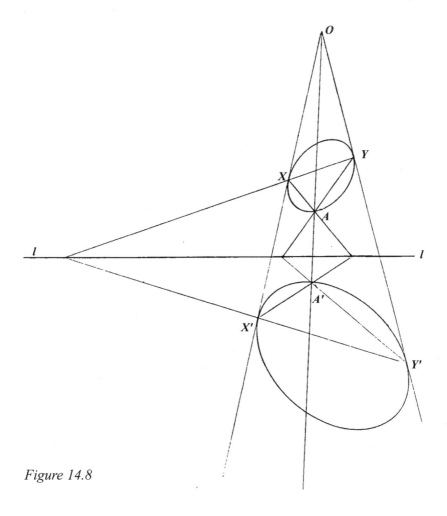

Figure 14.8

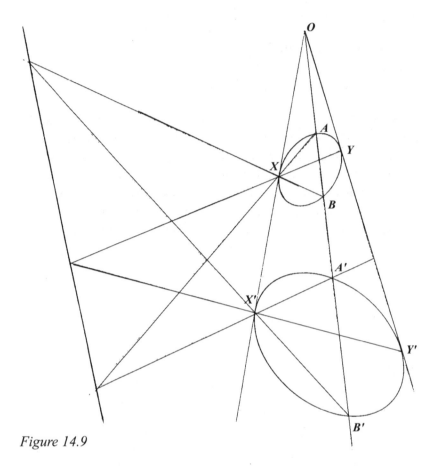

Figure 14.9

will be able to assure himself that these two ellipses really do meet along this line.

But it would have been possible to letter our line *OA* differently, as in Figure 14.9. This will give us a new homology, and a new line *l*, with two further common points of the two ellipses. Try it and see! And now we have found our four common points. But what about the six common chords? Well, we have two of them: our two lines *l*. The other four are imaginary lines, and when we have mastered Chapter 18 we will be able to find them also.

Exercise 14d
Try to dualize the contents of this chapter.

15. The Imaginary Part of a Conic

15.1 Representing imaginary points by involutions

The concept of an infinite group of circling transformations is a rich and rewarding one, and it is surely very near to the truth of what we are trying to study. But it is a cumbersome one to employ. We need a simpler method of expressing it if we are to work with any facility. And we find this in the 2-cycle, the involution. If we study the pairing of the mates in a circling involution, we have in fact before us the essence of the transformation-movements of that whole group of transformations. Thus we can take the involution to represent all the movement-relationships involved in the pair of imaginary points in question.

Now we know that an involution is determined by any two pairs. Thus we have only to put down two pairs of points (separating one another so as to ensure that the involution is a circling one) and this may be taken to represent to us not only an involution, but also a pair of imaginary points (Figure 15.1).

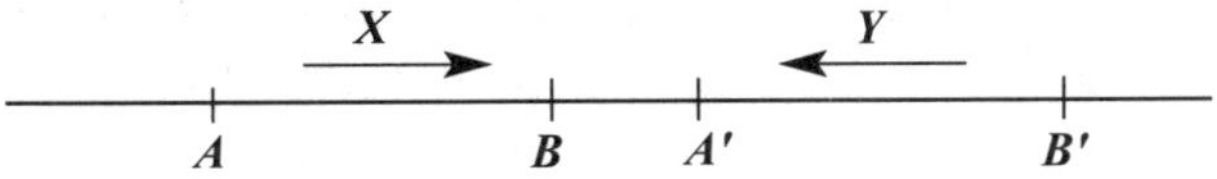

Figure 15.1

These two pairs determine an involution — and this involution may be taken to represent to us the totality of transformation movements possible with the same centre and the same amplitude. Thus they represent a pair of imaginary points, point X being the sum total of movements from left to right, and point Y being those from right to left.

But there are two pairs in the involution which are more convenient, and more significant, to take than any other pairs. These are the centre and its mate, the point at infinity, and the pair which is harmonic with this. This will naturally be the pair which is equidistant from the centre, and it will be the pair which is formed by the lines of the pencil in O which bisect those going to the centre and the point at infinity. These are the same points representing the centre and amplitude in Figure 13.3.

Having decided that we can let the involution represent for us all the transformations of the group to which it belongs we shall now show that the results of Chapters 13 and 14 can be reached by rather different means. The basic ideas are still the same, but because we are concentrating simply on the involutions we have the advantage of considerably simpler working; however we must realize that this simplification has been bought at the cost of impoverishing our concepts; this will not matter as long as we carefully remember, all the time we are working, that each involution must be taken to stand for all the infinite richness of the group of transformations to which it belongs. We shall be relying heavily on the work of Chapter 9 and the reader should be sure that he has a good memory of all that we learnt there.

15.2 The meet of a conic and a line

Let us consider any conic, and a line l meeting it in two real points X and Y (Figure 15.2). On this line take any point, A, outside the conic. From A we know there are just two tangents. If we draw the line connecting the points of contact of these tangents with the conic, we shall arrive at the polar, a, of point A (Section 10.1). If we take several further points of the line l, B, C, D etc, we can similarly arrive at their polars, b, c, d etc. Now we know from the theory of pole and polar, that the points A, B... are harmonic with the points in which their respective polars meet line l. In other words, these pairs of points are pairs in involution and it is clear from the figure that X and Y are the double points of this involution.

We may say this in a more general way. Any conic, when confronted by a line which meets it in real points, determines on that line an invo-

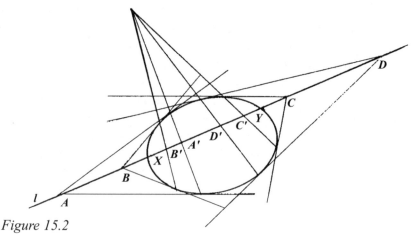

Figure 15.2

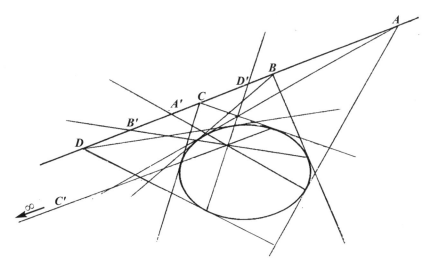

Figure 15.3

lution the double points of which are the points in which the line meets
the conic.

But we can try the same thing with a line which does not meet the
conic in real points. In Figure 15.3 we have made exactly the same
construction and we find that the conic has now made a circling involution on the line. But we know that this circling involution represents
transformation movements, from left to right, and from right to left,
which are a pair of imaginary points, the double, or invariant, points of
the involution. We have thus found the pair of points in which the
conic meets the line; they are an imaginary pair, and they are just as
much points of the conic as those which we actually see lying on the
visible curve.

Suppose we have a conic and a line m (Figure 15.4). We wish to
construct the amplitude of the imaginary pair in which the conic meets
the line. First we must find the centre of the involution. We take the
point at infinity of m, and draw the two tangents from it to the conic
(parallel to m). We join their points of contact, and this line gives us
point O, the centre of involution.

Now we refer to Figure 6.20 (page 90). There we showed, following from a consideration of Pascal's Theorem, some remarkable facts
regarding any four points lying on a conic. Notice, in this figure, that
UD, passing through A and D, is the polar of E, and similarly UC is the
polar of G. It follows that EG is the polar of U, and, therefore, that U

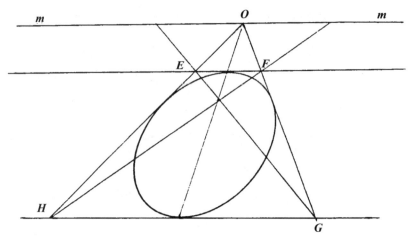

Figure 15.4

and *S* are mates in the involution that the conic determines on line *STUV*. Further, from the qualities of the quadrangle *EFGH* it is clear that *S* and *U* are harmonic with respect to *T* and *V*.

Hence the construction for the amplitude which a conic makes on any given line can be done as follows.

Having found the centre, *O*, we draw the tangents from this centre to the conic. Let these tangents meet the tangents which are parallel to *m* at *E*, *F*, *G* and *H*. *FH* and *EG* will meet the line *m* in two points which are a pair of mates in the required involution, and they are obviously the only pair which can be harmonic with *O* and the point at infinity. These two points are, therefore, the amplitude we are seeking.

15.3 The imaginary part of a circle

Now let us construct the amplitudes which a circle determines on a set of parallel lines, using this construction. You will find that these amplitudes lie on a rectangular hyperbola (Figure 15.5). The circle reveals itself as an organism which extends all the way to infinity and back! We must guard against the error of thinking that these points of the amplitude are the imaginary points — or that the imaginary part of a circle is a rectangular hyperbola. The amplitude simply marks for us the 'breadth' of the movement connected with the circle on that line. We might say that the circle 'breathes wider' as the line moves away from its real part, or, perhaps, that it moves farther into the imaginary. As the line approaches the real part of the curve, the

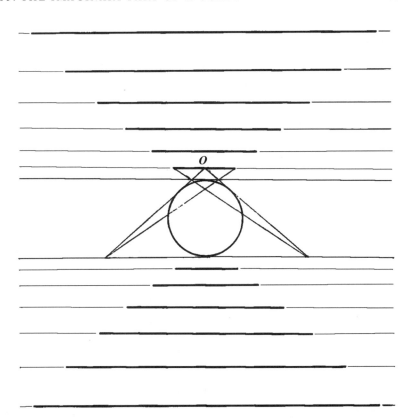

Figure 15.5

amplitude shrinks to a very small width — all the rest of the line finds its corresponding points within this tiny section and one might almost speak of a 'concentration of virtue' forming in this little space — finally the line moves in until it is a tangent, the amplitude becomes nil, the two imaginary points have come together into one, and they appear as a real point visible on the curve — the double point of contact of a tangent with its curve.

Exercise 15a

Repeat the above construction using first an ellipse, and then a hyperbola. Note that the amplitudes for the ellipse form a hyperbola (*not rectangular*) and that this touches its ellipse in a most harmonious and satisfying way. Note the curious interchange of function between the circle and the rectangular hyperbola. Each represents the imaginary part of the other.

15.4 Representation of imaginary lines as involutions

Just as a line, when considered within a plane, is dual to a point, so we can dualize the concepts of this chapter, and can come to the idea of a pair of imaginary lines. These would be represented by a circling involution of lines in a point. If we put down any two pairs of lines through a point P, the pairs to separate one another, then we have determined a circling involution in P. This involution will represent to us all the transformation movements of lines in P which have a certain pair of imaginary lines as their invariant lines. The two sets of movements, clockwise and anti-clockwise, taken in their totalities, are each an imaginary line of the conjugate pair. A pair of conjugate imaginary lines have a real common point P, in the same way that a pair of conjugate imaginary points share a real common line — the line that bears their transformation movements.

We refer back to Section 9.3. It was shown there that in any circling

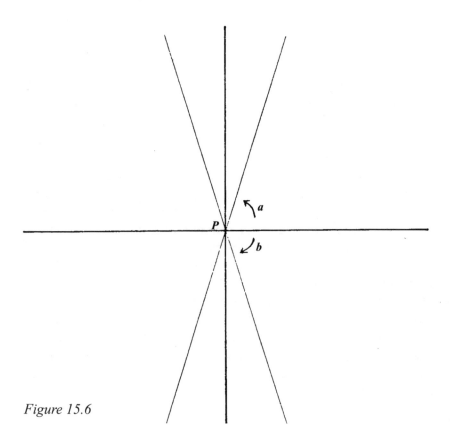

Figure 15.6

involution of lines there is either just one pair which is right-angled, or, in one special case, all pairs are so related. The amplitude of a pair of imaginary lines is set by the pair of mates which is harmonic with the right-angled pair. From what was shown in Section 9.3, it follows that these lines are bisected by the right-angled pair. We can now make a representation of a pair of imaginary lines, *a* and *b* (Figure 15.6).

It is important that the reader should do the following exercise, as the understanding of what follows depends upon it.

Exercise 15b

Dualize, at least, Section 15.3. For this we need to start by finding the dual of the point at infinity of our base line *m*. The most natural position for this is a line through our base-point *M*, which also passes through the centre of the circle, the centre and the infinite being normally polar to one another. You will find that you arrive at a rectangular hyperbola, constructed linewise, each pair of tangents in the construction meeting at a point within the circle, and forming the amplitude of a pair of imaginary lines, which are the tangents to the circle at that point. We begin to see what a complex and wonderful organism a circle really is!

15.5 The absolute circling points

Let us consider once again the picture of the imaginary part of a circle, Figure 15.5. If we wish to find a pair of mates on any line of the parallel set, we simply take the points at which that line meets the hyperbola. The arms of this hyperbola, as they widen towards infinity mark the amplitudes on ever more and more distant lines. Now the asymptotes of such a hyperbola meet the line at infinity in points which are 90° apart, therefore, we may say that these points are a pair of right-angled points on the infinite line as mates in this involution. It follows, therefore, that the circle determines on the line at infinity an involution which is completely right-angled. But it is also obvious that every other circle will do the same.

We now have a clue to a problem which might otherwise vex us. We know (Section 12.1) that two conics may meet in four points. Circles are conics, but we know that two circles can only meet in two points. To be more accurate we should say they can only meet in two real points, and now we can see the reason. All circles, by virtue of the very fact that they are circles, share these two imaginary points on the infinite line in common. These two imaginary points, in which all circles meet, are often called the *absolute circling points at infinity*, and they are usually denoted by the letters I and J. As soon as we go to put two conics on a page, and we say 'These shall be circles,' we are determining that they shall share in common (meet in) the points I and J. Two distinct conics cannot meet in more than four points, therefore there are only two other points in which they can meet. These may be real or imaginary. Of course this right-angled involution in which every circle meets the infinite line stands for, or represents to us, a whole group of circling measures. Each of these circling measures marks out equal spacing along the infinite line. This is a most important fact to grasp, and remember. Whenever we meet a series of points along the line at infinity which are equally spaced, they are members of a circling measure which moves between the points I and J as imaginary invariant points.

15.6 Measuring distance and angle

We now have the concepts necessary to transform our general projective plane into a metrical one. In a projective plane we have points and lines, and we study the incidence-relationship between them, but we have no notions of size, distance or parallelism — in fact, we have no possibility of measurement. If we are to measure, we must have some fixed standard from which to measure. In a Euclidean-type plane (there are others) it is found that we need to choose one line from the infinitude available to us, and any two points on that line, and to regard these as fixed; these we call our *absolute elements*, and from them we make all our measurements. In the Euclidean plane the absolute line is the line at infinity and the absolute points are I and J.

We can now describe some of our metrical properties in projective terms.
1) Two lines which meet on the absolute line are *parallel* (we have had this one before.)
2) Any two lines which meet the absolute line in points which are harmonic with respect to the absolute points are *right-angled*.
3) Any step measure which has its double point on the absolute line marks *equal intervals*.
4) Any growth measure which has one of its double points on the absolute line marks a *geometric series*. (We have had 3 and 4 before).
5) Any conic which passes through the absolute points is a *circle*.

6) The pole of the absolute line with respect to any conic is its *centre*.

7) Any two lines from this centre to the conic, must be considered as having equal *length* (the projective principle of *rotation*).

Now let us construct a Euclidean-type plane, different from the actual Euclidean plane. Our absolute line shall be *z* and we choose two real points, *I'* and *J'* on it to be our absolute points (Figure 15.7). Through them we draw a conic. We draw the tangents to the conic where it meets *z*, and their common point, *O*, is the pole of *z* and therefore, the centre of the conic. Now we draw any line through *O*, to meet the conic in *S* and *T*. This is a diameter. Now we choose any further point, *U*, of the conic and we join *SU* and *TU*. Now we know that *S* and *T* are mates in an involution of which *I'* and *J'* are the double points. Therefore, *S I' T J'* are harmonic, and it follows that the lines *UT*, *UI'*, *US* and *UJ'* are harmonic.

Now repeat this with

— *z* being the line at infinity

— *I'*=*I* and *J'*=*J*

and thus the conic being a circle. The lines *UT*, *UI*, *US* and *UJ* are harmonic, therefore, the lines *UT* and *US* are at right angles. From purely projective consideration we have arrived at the fact that the angle in a semi-circle is a right angle! Many of the other Euclidean theorems about circles can similarly be seen to be simply metrical cases of much more general projective truths.

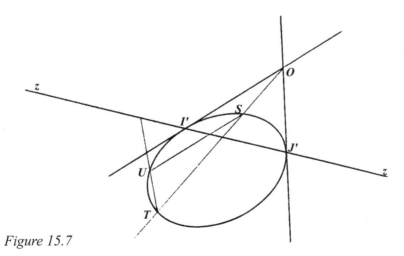

Figure 15.7

It is a very instructive exercise to try this with various well-known Euclidean theorems. Each time one does it one reaches a figure in one's quasi-Euclidean plane which looks quite different from the original Euclidean figure but which when interpreted in terms of that quasi-Euclidean plane, is identical with the original theorem. But if one now regards one's new figure in terms of the ordinary Euclidean plane, one finds that one has reached a new, usually projective, theorem, which can then be verified in terms of ordinary projective geometry.

Exercise 15c

Euclidean theorems which it is useful to try this with, include the following:

1) At any point on a circle the tangent and the radius are perpendicular.
2) Angles in a segment of a circle are equal.
3) The line from the centre of a circle to the midpoint of a chord is perpendicular to it.
4) The line from the apex of an isosceles triangle perpendicular to the base, meets it at its midpoint. For the purpose of this exercise, construct your isosceles triangle using rule 7 of the previous section (page 171).

16. Four-Point Systems with Imaginary Elements

16.1 Orders of magnitude

Before going any further we must clarify one or two ideas about the imaginary points and lines.

Firstly, let us consider the real points of a plane. In any line of the plane we have an infinitude of points. If we now let this line turn round one of its points, we will have an infinity of lines, and a moment's reflection will show us that these lines will cover all the possible real points of the plane. The pencil, therefore, contains an infinity of lines, and each line contains an infinity of points. It is a consideration such as this that leads us to say that the plane contains ∞^2 points. By dualizing the above process we can easily see that the plane also contains ∞^2 lines. When speaking of ∞^1, ∞^2, ∞^3 etc., we must beware of falling into the habit of thinking of these as though they were numbers or quantities; to do so is to be led into impossible paradoxes. When one moves from ∞^1 to ∞^2 it is more true to say that one is moving into a new *order of magnitude* — a new realm of magnitude altogether. A line, in its infinite extent, is far larger than an ellipse, yet we say that they both contain ∞^1 points, and we can establish a one-to-one relation between the points of the one and of the other. But we can never have a one-to-one projective relationship between the points of a line and of a plane: ∞^1 and ∞^2 are different orders of magnitude.*

Now imagine any line. We can take any point of it which we wish for the centre of a circling involution, and with this centre there are an infinite number of possible amplitudes. We can, therefore, have an infinite number of imaginary points associated with this centre (we must accept here the idea that two times ∞ equals ∞). But there is an infinity of possible centres which we could have chosen. It follows,

---

* The exponents in these powers of represent the *degree of freedom* of the concerning element bound by two others. So the degree of freedom of a point in a plane equals 2: it is bound only by this plane, and hence also by the Nothing. The degree of freedom of a line in space through some point equals 2; it is bound by this point and Space.

therefore, that while a line contains ∞^1 real points it contains ∞^2 imaginary ones.

Similarly, consider our picture of the imaginary part of a circle or an ellipse. Along a series of parallel lines we found ∞^1 imaginary points. But there were ∞^1 different sets of parallel lines which we could have chosen — one for each direction in the plane — and it follows from this that whereas the real part of a circle or an ellipse contains only ∞^1 points, the imaginary part contains ∞^2 points. In each case the unseen part reveals itself as being infinity times richer than the seen part.

16.2 The imaginary triangle

This leads us to another important consideration. A line can be considered in two different ways. Either we see it as an entity in itself, or we see it as being composed of an infinite number of points. Dually, a point is an entity in itself or it is composed of an infinite number of lines. These two cases are exactly similar; it is only the limitations of our own consciousness which make the first so much easier to see than the second.

Now we put on the page an imaginary line, shown by the right-angled pair of its involution and the pair of mates harmonic with (bisected by) it. We will call the line m the clockwise rotation (Figure 16.1).

Now we ask 'If we consider this line to be a line of points, where are all these points?' The answer is a very simple one. Put any real line l on the page. The given involution will project into a circling involution of points on line l and this involution represents an imaginary point on l, say point M. M is one of the imaginary points making up line m. Notice that we could have put any one of 2 lines of the plane instead of l, and on each of these the involution of line m would project an imaginary point. We see, therefore, that an imaginary line contains 2 imaginary points just as a real line does.

Notice that in Figure 16.2 we have constructed an *imaginary triangle*. Line m has a conjugate imaginary line n which is the anti-clockwise motion. Lines m and n meet in point L (a real point). n projects onto real line l in imaginary point N. Our triangle has, therefore, two imaginary lines, m and n, and one real line, l. We shall often have to work with such triangles.

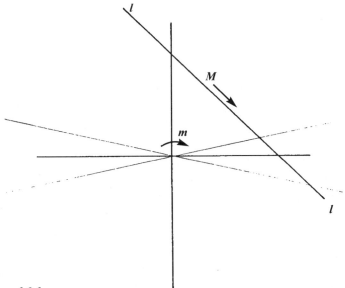

Figure 16.1

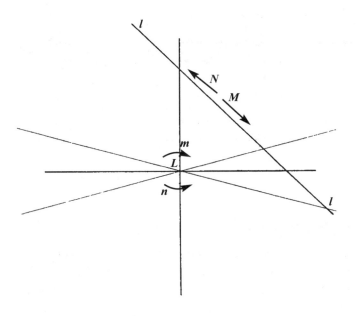

Figure 16.2

16.3 A self polar triangle on a circle

Now let us revert to the picture of the imaginary part of a circle (Section 15.3). When we do this pointwise we take a line, *a*, outside the circle and we find the amplitude of the imaginary points, *B* and *C*, in which the circle meets this line (Figure 16.3). The points of this amplitude we know, lie on a rectangular hyperbola. Now we ask, 'If the imaginary points *B* and *C* lie on the circle there must be imaginary tangents which touch the circle in these points where are they?' To find them we first find point *A*, the polar of line *a* with respect to the circle. If we now construct the imaginary lines (tangents) of the circle through point *A* we will find that the amplitude of this involution, represented by lines *s* and *t*, is in exact perspective with the involution of points *B* and *C*, and we have constructed a *self-polar triangle* on our circle. The picture as we have envisaged it is Figure 16.3 and its real analogue is in Figure 16.4. It is a *most* valuable exercise to construct the left-hand picture for yourself according to the constructions given in Figure 15.5 and Exercise 15b (pages 167, 169).

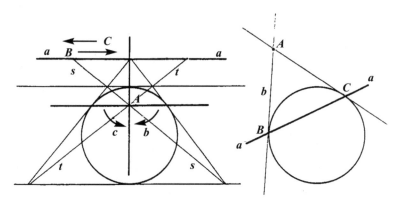

Figure 16.3   Figure 16.4

16.4 Concentric circles

Now consider a circle and the line at infinity. We know that the circle meets the infinite line in the points *I* and *J*. From the pole of the infinite line (that is, the centre of the circle) there will be imaginary tangents which will be in direct perspective with *I* and *J*. These tangents, *i* and *j*, are represented by right-angled circling involutions at the centre, *O*, of the circle (Figure 16.5). But another circle with the same

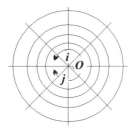

Figure 16.5

centre O would have the same tangents i and j, meeting it in the same points, I and J.

We can now draw a set of concentric circles and its analogue, when the line o has become accessible, and the points I and J have become the real points I' and J' (Figure16.6).

Concentric circles now reveal themselves to us as a pencil of four-point conics, having two coincident points in common at I and two at J. In other words, these circles all meet on the degenerate 4-point $IIJJ$. Where are the six lines of this four-point?

To answer this let us draw one of the conics of Figure 16.6 with the lines OI' and OJ' not quite tangent (Figure 16.7). We can then easily see the six lines we are looking for.

An examination of this shows us that the six lines in Figure 16.6 are $I'J'$ counted four times and OI' and OJ'.

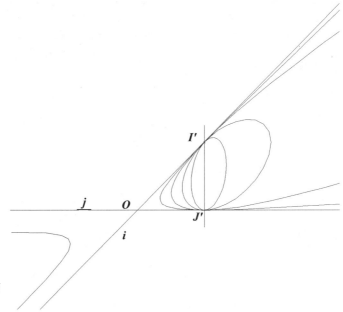

Figure 16.6

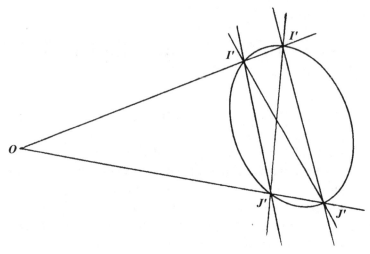

Figure 16.7

We now cut the concentric circles by any line *m* tangent to one of
the circles at, say, *X*. We know from Desargue's Conic Theorem
(Section 12.2) that the circles will meet this line in pairs of points in
involution, seeing that we know that the circles are, in fact, a system
of four-point conics. One of the double points of the involution will be
the point at infinity of *m*, *Y*, since the degenerate conic *IJIJ* meets it in
two coincident points there. The other double point will obviously be
at *X*. Every other circle of the system will meet the line in a pair of
points, which are harmonic with *X* and *Y*, or in other words, which are
equidistant from *X*, since *Y* is at infinity. But this is just one of the obvious things with which we are well familiar when dealing with concentric circles, so we see that the projective truths hold good even when
an essential part of the figure is imaginary and, therefore, invisible.

17. The Projective Idea of a Circle

17.1 Constructing a circle projectively

Although the circle is not usually considered a projective curve, as its ordinary definition involves the idea of distance, we now have a means of realizing the true projective nature of it: a circle is a conic which passes through (contains) the points I and J.

With this fact in mind we now arrive at a truly projective construction of a circle. We start with a well-known Euclidean theorem: the angle in a semi-circle is a right angle.

If A and B are the ends of a diameter, and P is any point on the curve, then the angle APB is necessarily a right angle (Figure 17.1).

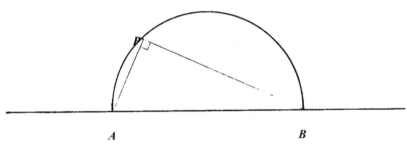

A B

Figure 17.1

This is the Euclidean, and partial, view of the picture. Projectively, we notice that the lines AP and BP meet the line at infinity in points that are 90° apart, or in other words, which are mates in the involution of which I and J are the double points. Any range of points on a line m is projective with the range formed by the harmonic mates on m of those points. Thus as P moves round the circle there are formed projective pencils, which is the fundamental projective way of generating a conic (Section 6.1). Once we have the I–J involution on the infinite line we can construct the circle without any need of compasses.

17.2 Concentric circles in perspective

This leads to an extremely important construction. We will do
exactly the same thing, but this time have the absolute ('infinite')
line on the page. Let *z* be our absolute line (Figure 17.2). On it we
put a pair of imaginary points, I' and J'. Instead of marking the
amplitude of their involution we will mark some equiangular, cyclic,
circling measure belonging to the measures which I' and J' determine on that line. This is just as sure as, though usually less convenient than, marking their involution. For convenience, we will
forsake our projective principles and use a protractor! At some convenient point near the line we make an equi-angular pencil, the lines
of which shall be, say, $10°$ apart, and we ensure that one of these

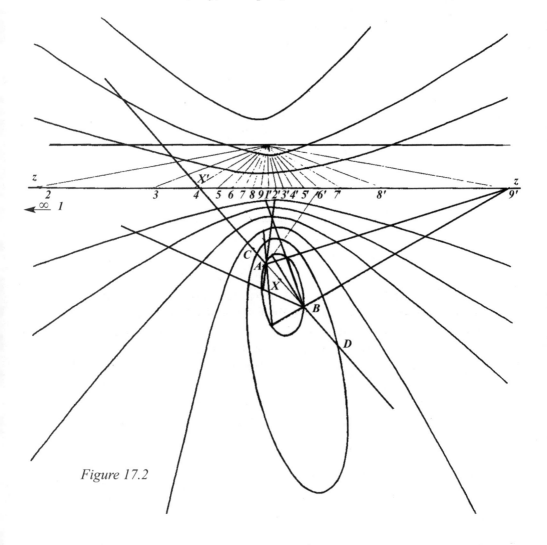

Figure 17.2

lines is parallel to z. These lines project into the points of z which form the circling measure we want. We say that these points are separated from one another by $10°$ within the circling measure of I' and J'. Now we number the points *1, 2, 3* ... up to *9*, starting from the point at infinity, and then continue with I' $2'$ $3'$ ... to $9'$. We will now find that the points *1* and *1'*, *2* and *2'* etc, are separated by $90°$. Next we take any two points A and B, and let the line AB meet z in point X'. Find X, harmonic to X' with respect to A and B. Point X will be the 'centre' of our 'circle'. Now draw, very lightly, lines from A to point *1* on z and from B to *1'*. These count as lines at right angles, and they therefore meet on the 'circle.' Now continue with joining $A - 2$, $B - 2'$ etc. until you come to $A - 1'$ and $B - 1$, etc. When you have finished you have drawn a conic which, with respect to line z and points I' and J' as absolute elements, is undoubtedly a circle. All the qualities which an ordinary circle bears to the line at infinity, this conic bears to line z.

By taking another pair of points, C and D, harmonic with respect to X and X' we can find another 'circle' concentric with the first. By continuing this way we can construct a whole family of concentric 'circles.' These are, in fact, conics, varying through different forms of ellipse, through one parabola and then through various forms of hyperbola. In the hyperbolas we can see how they gradually straighten out onto line z, a true perspective picture of the way in which a circle, as it becomes of infinite radius, straightens onto the line at infinity. We see clearly that as the hyperbola does this it approaches the line z from two sides at once, and that finally it must lie doubly upon it. This is a reminder that as a circle becomes of infinite radius and degenerates into the line at infinity, this line, considered as a degenerate circle, has to be counted twice, thus preserving the order of the curve.

These conics all meet line z in the same pair of imaginary points; they all have the same pair of imaginary tangents through point X; and they all make point X and line z pole and polar to one another. They are, in fact, projectively equivalent to Figure 16.5.*

---

* These curves form a system of path curves. See Section 21.10.

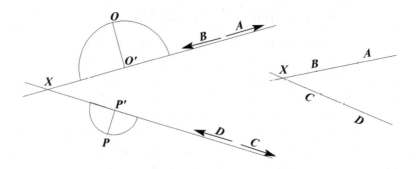

Figure 17.6   Figure 17.4

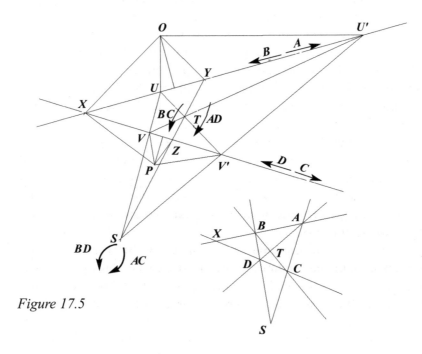

Figure 17.5

Figure 17.6

17.3 Joining imaginary points and lines

Consider any two lines on the plane, and a pair of conjugate imaginary points in each, *A B* and *C D*, represented by their centres (*O'* and *P'*) and amplitudes (the diameters of the semi- circles) in Figure 17.3. This gives us a quadrangle of four points; we know that through four points there pass six lines, and it is now our aim to find the six lines in this case. Two of them are obvious; they are the real line-carriers of the imaginary pairs. We give the real analogue in Figure 17.4.

Apart from the pair of lines *AB* and *CD*, which we already have, we have to look for the pairs *AC* and *BD*, and *AD* and *BC*. These will be imaginary pairs represented by involutions of lines in two points and these points will have to be so placed that they are in direct perspective with the circling measures on *AB* and *CD*.

Now we know the conditions for a direct perspective along these two lines: it is that the common point of the two lines must be self-corresponding (Section 5.2). Therefore, we will start with this point, *X*. First construct the mate of *X* in each of the two involutions; we will call these *Y* and *Z*. The method for this is to draw a semi-circle on the amplitude of *AB* and let this be met in point *O* by a perpendicular through the centre point of the involution (Figure 17.3). The mate of *X* will subtend a right angle with *X*, as seen from *O*, *Y* in Figure 17.5. Next construct the unique pair on *AB* which is harmonic with *XY*. We do this by taking a line line through the centre, *O*, at 45° to the lines *OX* and *OY*. And a similar construction in *P*. We call these points *U* and *U'* on the one line, and *V* and *V'* on the other. Now it is clear that the points *XUYU'* on the one line are projective with *XVZV'* on the other, seeing that they both form harmonic ranges. Their common point is self-corresponding, therefore, they are in direct perspective and the lines *UV*, *YZ* and *U'V'* will be concurrent. But since these are harmonic ranges we can interchange the functions of *U* and *U'* without altering the cross ratio (Section 4.1). Therefore we will also find that the three lines *UV'*, *YZ* and *VU'* are also concurrent. We will call these points of concurrency *S* and *T*.

Now we have our four other lines — an imaginary pair in *S* and another in *T*. From both *S* and *T* the two involutions are completely in direct perspective. We label the lines as in the figure, and you will easily see how they correspond with the real analogue (Figure 17.6).

Here we have constructed the complete imaginary 4-point, consisting of 4 imaginary points, 2 real and four imaginary lines, and a real diagonal triangle, *XST*. All the harmonic relationships of the real four-point are to be found also in the imaginary four-point. Consider the line *XT* in the real four-point. We know (Section 4.3) that this line will be met by *BD* and *AC* in two points which are harmonic with respect to *X* and *T*. Now in the imaginary case *BD* and *AC* will project an involution onto line *XT* from point *S*. Notice that this involution is in perspective from *S* with the *C–D* one, and that *X* and *T* are projections of *X* and *Z*. But *X* and *Z* are mated in the *C–D* involution therefore *X* and *T* must be mates in the involution along line *XT*. This being so we know that they are harmonic with the imaginary points of that line, since these are the double points of the involution, and any pair of mates must be harmonic with the double points (Section 9.1). Therefore, *BD* and *AC* meet line *XT* in points harmonic with respect to *X* and *T*.

Exercise 17a

It is most valuable to draw Figure 17.5 for oneself, marking in the rotation directions of the lines *AC*, *BD*, *BC* and *AD* without referring to those already marked in the figure given here. It will be found to be much simpler than it looks when it is put complete before one. Having done this, confirm, either by drawing or by the ordinary laws of perspective and projectivity, that line *ST* is in fact met by lines *CD* and *AB* in a pair of points harmonic with *S* and *T*, and that line *XT* is met in a pair harmonic with *X* and *T* by the lines *AC* and *BD*.

Having done all this you will have had some valuable experience in handling imaginary points and lines, and as you gradually find, one by one, all the projective properties of the real analogue appearing in the imaginary figure, you begin to have real confidence that you are, in fact, truly dealing with lines and points when you manipulate the involutions in this way.

Exercise 17b

Dualize Section 17.3: find the meeting point(s) of two (or four) imaginary lines.

18. The Imaginary Circle

18.1 To construct an imaginary circle

In Figure 17.1 we constructed a circle as the meets of right-angled lines through A and B. Now we are going to construct an imaginary analogue of this. We will repeat Figure 17.1 of Section 17.1, but this time using a pair of conjugate imaginary points for A and B, having their amplitude marked by points M and N (Figure 18.1). Firstly, we must put a line, l, through A. It will have to be an imaginary line and will therefore be an involution of lines through some point X. This point could be anywhere in the plane, outside line AB, but simply for purposes of convenience, we will place it somewhere along the perpendicular bisector of the amplitude of the pair AB. Figure 18.2 depicts real analogue.

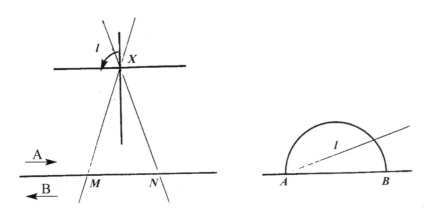

Figure 18.1   Figure 18.2

Now we must note where line l meets the line at infinity. It will do so in the imaginary point, C, represented by the involution on the infinite line determined by the four points in which the lines through X meet that line. If we turn all these four points through 90° we shall get the point, D, which we could reasonably consider as the mate of C in the I–J involution, and therefore at 'right-angles' to it. This point

D must be joined to *B*. If it is joined to *B* the lines from these new four points must go through the involutary pairs of point *B*; that is, through *M* and *N*, and through the centre and the infinite point of the *B*-involution. It will be easily seen that they must meet somewhere along the vertical line through *X*. We call their meet *X'*, and this will be the real carrier of the imaginary line through *B*, and we call this line *l'*, Figure 18.3. Now we have to find the point in which *l* and *l'* meet; in other words we have to find a line carrying an involution which is in direct perspective with the involutions in *X* and in *X'*. This is clearly the line *ST*, and *S* and *T* mark the amplitude of the point we are seeking. This then is the point *P* which lies on (is a point of) the circle.

Obviously, had we started with joining point *B* to *X* by line *k*, etc, we should have ended with point *P'* conjugate, on line *ST*, with point *P*. We have thus found the pair of imaginary points in which our circle meets line *ST*.

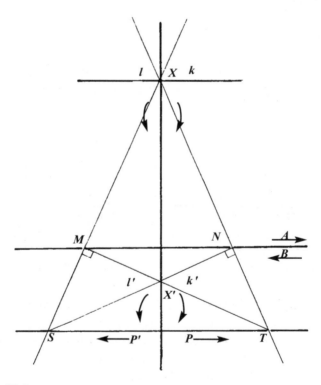

Figure 18.3

Exercise 18a

Draw a real analogue for Figure 18.3, remembering that points M and N do not lie on the circle but are simply some pair of points harmonic to A and B; also remembering that neither does S nor T lie on the circle.

It will easily be found that if we had taken point X at some other position along the perpendicular bisector of the amplitude of points A and B the construction would have given a pair of imaginary points of the circle along some other line parallel to AB, and with some other amplitude. We now wish to find the locus of points S and T as the line carrying them moves parallel to line AB.

Notice that as X moves along the vertical line, X' moves in sympathy, so that they always subtend a right angle from points N and M. They are thus mates in circling involution, and therefore mark out projective ranges. Therefore the lines MS and NS are lines of projective pencils and point S moves along a conic. Similarly of course, for point T. Clearly if the line MX is at 45° to MN, the line NX' will also be at this angle. Therefore the curve meets the line at infinity in the two points which are 45° away from the line MN. The curve is thus a rectangular hyperbola through M and N.

Figure 18.4 is a picture of all the amplitudes of the points in which an imaginary circle meets a pencil of parallel lines.

Now let us see how much we can find out about the conic we are constructing. In the real analogue of Figure 18.2 we can easily see how to find the centre of the circle — it is the harmonic mate of the point at infinity with respect to A and B. In a circling involution we know that the harmonic mate of infinity is the midpoint of the amplitude, so we can immediately mark in the centre, O, in Figure 18.4.

Now refer back carefully to Figure 15.5. We see there, how as a line approaches the real part of the circle, the amplitude gets gradually less, until at last it becomes nil: the two imaginary points have come together and in that moment have become the coincident real pair in which a tangent meets the curve. We see how the amplitude is in some way related to the distance apart of the imaginary points along their common line. Now if we wish to construct further points of our curve of the imaginary circle we must put another line through the centre O, inclined at any angle we wish to the line AB (MN) and on it we must put an involution with the same amplitude as MN, so that we are working with the same radius of circle. It we do this we shall obviously get

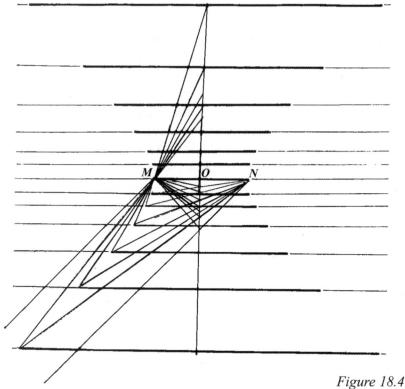

Figure 18.4

another rectangular hyperbola, with the same centre, and of the same size, as the other one, but simply inclined at some angle to it. So we must complete Figure 18.4 in our imagination by putting an infinitude of rectangular hyperbolas all centred at O and all tangent to the real circle through M and N.* Since these hyperbolas are all rectangular it is clear that this conic meets the line at infinity in I and J; therefore it can truly be called a circle.

Next notice that in Figure 15.5, the involutions are so arranged that there are two lines on which the amplitudes have become nil, and that between these two there is a space in which the lines carry breathing involutions, the double points of which are the real points of the circle. But in Figure 18.4 the amplitudes never get less than the distance M to N and therefore there can be no lines with real points. This is truly an imaginary circle; that is, one containing only imaginary points and no real ones.

---

* Note that the imaginary circle has ∞^2 points.

18.2 The real projection

Now we must find the radius of our imaginary circle. Notice that the equation of a circle is

$$x^2 + y^2 = r^2$$

where r is the radius. If $r^2 = 9$ we have a circle of radius 3, if it is 4 we have a circle of radius 2, until when r becomes zero the circle has shrunk to nil radius — it has become just a point. If r^2 becomes -1 the circle re-appears as an imaginary one, and gradually grows larger as r^2 becomes -4, -9, etc. With the real circles we have radii of 3, 2, 1 and then 0; we are tempted then to say that the radii of the imaginary circles will continue in this process and will have radii of -1, -2, -3 etc. But this is not a true way of looking at it. Where $r^2 = 9$ we know that $r = \pm 3$. We can see clearly how the radius of a real circle is plus or minus, from Figure 18.5.

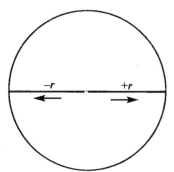

Figure 18.5

We measure the radius in one direction and call it plus; then measured in the opposite direction it must be minus. If a radius of -3 belongs to the real circle it cannot belong to the imaginary one. So again we ask: what is the radius of an imaginary circle. Clearly the radius of a circle with an equation

$$x^2 + y^2 = -1$$

must be $\sqrt{-1}$ which is i. If $r^2 = -9$ the radius would be $3i$ (3 times i).

In fact, the radius of our imaginary circle is i times half the distance *MN*. This distance, the semi-amplitude along the diameter, is called the *(real) conjugate radius* of the imaginary circle; and the circle, centre *O*, through *M* and *N* is called its *(real) conjugate circle*. We may say that it is what the circle would have looked like if it had been real.

18.3 Pole and polar with respect to an imaginary circle

An imaginary circle contains all the relationship-possibilities of a real circle, such as, for instance, the pole and polar laws. How do we find the polar of a point with respect to a real circle?

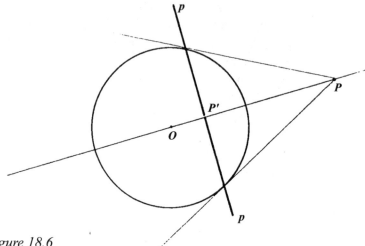

Figure 18.6

We take the point P and join it to the centre, Figure 18.6. We find point P' harmonic to P with respect to the points in which the line meets the circle. Through this we draw the perpendicular and this is the polar line we are seeking, p. As P moves inward towards the real part of the circle, p moves outwards towards it. When P reaches the circle and becomes a point of it, p becomes the tangent at that point. The two move in reciprocal, breathing movement. In the moment in which they become incident one with another the real part of the circle is generated.

The right-angled relation of the polar p to the line OP is clearly connected with the fact that the circle contains the points I and J. We shall find the same relation with the imaginary circle, but in this case the movement is a circling one. Let $MO = ON$ be the conjugate radius of an imaginary circle, Figure 18.7. We wish to find the polar line of a point P.

Draw the diameter-line OP, and find the harmonic mate of P with respect to the points in which the imaginary circle meets OP. That is to say, the harmonic mate of P in the circling involution along OP, centre O and amplitude equal to the distance MN. This is very easily done.

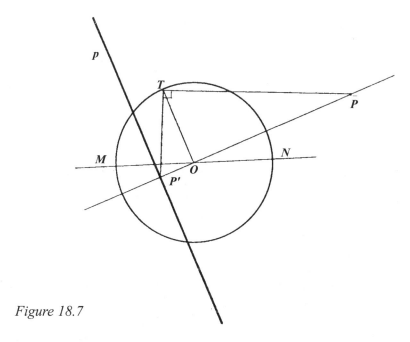

Figure 18.7

Draw the real conjugate circle. Find the end, *T*, of the diameter at right angles to *OP*, and draw the line at right angles to *PT*. This meets *OP* in *P'*, and the perpendicular to line *OP* through *P'* is the polar we are seeking, *p*.

As *P* moves outwards away from the centre, *p* moves inwards toward the centre. They move in circling measure, one chasing the other around the line. Thus they never come together, and the circle never has a chance of having any real points.

18.4 The meet of an imaginary circle and a line

By using a little metrical geometry we can find a more convenient way than the construction of Figure 18.3. We reproduce its lower part, with the midpoint of the *P–P'* involution marked *C*, Figure 18.8.

Through *C* we draw a line parallel to *TM*, and therefore at right angles to *MS*, meeting it at *D*. Since *C* is the midpoint of *ST*, by a well-known Euclidean theorem, *D* must be the midpoint of *MS*, and it therefore follows that *CM* = *CS*. In other words *CM* equals the semi-amplitude of the involution which the imaginary circle determines along the line *ST*.

Figure 18.9 will make it quite clear how to construct the amplitude

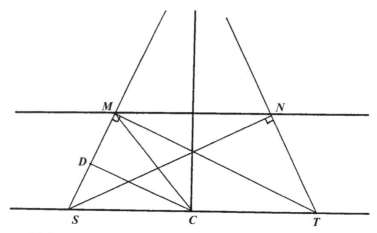

Figure. 18.8

of the imaginary points in which the imaginary circle, whose conjugate circle is the circle α, meets the line m. With centre C and radius CM make arcs at S and T. These mark the ends of the amplitude.

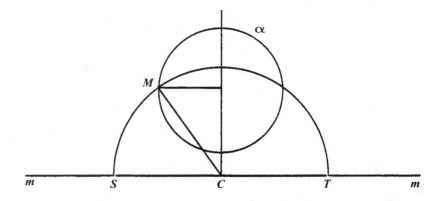

Figure. 18.9

19. Four-Point Systems of Conics

19.1 Conics through four imaginary points

We will start by repeating a Figure (19.1) which is already familiar to us — but for our present purpose we will letter it rather differently. And we are going to consider this system of conics passing through the points *I*, *C*, *J* and *D*. Now we will suppose that these four points all become imaginary and that the line *IXJY* becomes the line at infinity and the points *I* and *J* become the absolute imaginary circling points of our plane.

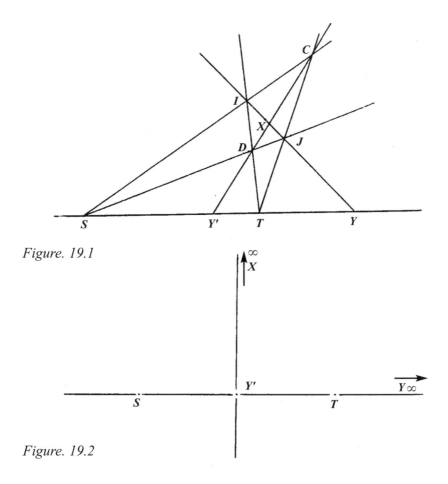

Figure. 19.1

Figure. 19.2

In Figure 19.2 we picture what the straight lines of our figure would then look like. We can place S and T anywhere we wish on our page; then since Y is at infinity, and Y' is harmonic to it with respect to to S and T, Y' must come at the midpoint of the segment ST. X is on the line at infinity, harmonic to Y with respect to I and J. X and Y are therefore a pair of harmonic mates in the I–J involution. and we know from what has gone before that they must therefore be separated by 90°. Therefore, we see that the line $Y'X$ must be perpendicular to line ST.

Next we imagine all the possible conics going through the points I, C, J and D. What can we find about them?

— Firstly, since they go through I and J they must all be circles.
— Secondly, since X is harmonic to Y' with respect to D and C, and to Y with respect to I and J, it follows that TY is polar to X with respect to every conic of the system; therefore every line through X must have its pole on TY; therefore the centre of each circle must lie somewhere along line TY.
— Thirdly, we know that the conics of the pencil will meet this line in pairs of points in involution, and it is obvious from the figure that the double points of this involution must be T and S. Therefore, we can see that all our circles will meet line TY in points that are harmonic with respect to T and S. This is easy to draw. It is the well-known system of Apollonian Circles (Figure 19.3).

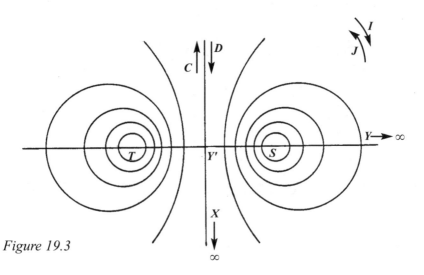

Figure 19.3

Exercise 19a

Construct such a system of Apollonian Circles. Then submit them to a homology, thus getting a general picture of a system of conics having four common imaginary points. Such a picture is beautiful, and can give one a strong feel of how the invisible generating elements can impart a sense of unity and harmony to a figure. See also Figure 19.4.

Figure 19.4

19.2 The quality of Apollonian circles

If these circles really form a system of conics through four common
points they should exhibit all the qualities of four-point systems which
we examined in Chapter 12, and it is good to test for some of these. It
is easy to see how it is that any line will touch just two of the circles.
Draw such a line, tangent to any two of them, and we know that the
points of contact must be the double points of the involution which the
system determines on that line. The perpendicular line through Y' (that
is the line DC), and the line through I and J (that is the line at infinity)
together form one of the degenerate conics of the system, and these
should therefore meet our line in points which are harmonic with
respect to the double points. In other words, the line DC should bisect
the distance between the points of contact on any line we like to
choose.

Next take any two or three of the circles and construct, by the
method of Section 15.2, the imaginary points which they have on line
DC. You will find that in every case the result is the same. All these
circles really do share two points in common along that line. Notice
that the amplitude of these common imaginaries subtends a right angle
at both S and T. In other words S and T are the meeting points of the
four lines joining D and C to I and J; this is the expression in this figure of the fact that, in Figure 19.1, IC and JD, ID and JC, pass through
S and T respectively.

There are two further things we can do to convince ourselves that
these circles really are in the relation of four-point conics. In Section
12.3, we learnt that if we take any general point of the plane, and find
its polars in the conics of such a system, they will form a pencil of
lines. Try this on your figure and confirm that this is so.

Next, remember what was said in the same section, that if we take
any general line of the plane, and find its poles with respect to the
conics of the system, these points will lie on a conic. Try this on
your figure, and an interesting result will appear. You will find that
you are beginning to construct what appears to be a conic running
into points S and T; only part of the curve appears. One now has to
ask where one can find the other part of the curve.

A moment's thought will show us that in fact our Apollonian
circles, as we have drawn them so far, represent only half the true
four-point system. We have asked for all the conics passing through
I, J, C and D; in other words, for all the circles passing through C

Figure 19.5

and D. We have considered all the real circles; but there will be an infinitude of imaginary ones as well. We cannot draw these but we can draw their conjugate circles quite easily. Refer back to the quick way of constructing the amplitude of the points which an imaginary circle contains on any given line (Section 18.4). We draw a circle, centre Y', and radius $Y'S$. This will also pass through the amplitude points of C and D, say M and N, and if we take any point of this circle, and drop a perpendicular on to line TS, the foot of this perpendicular will be the centre of one of our conjugate circles, and its length will be the radius. Figure 19.5 presents the picture of the conjugate circles of the imaginary circles passing through C and D.

These imaginary circles belong just as truly to the system of four-point conics which we call Apollonian circles, as do the real ones. Continue plotting in the poles of our general line, with respect to these imaginary circles and you will find that the remainder of the curve appears just as we should expect it. By working with processes like these we can begin to have confidence that when we work with imaginary elements, points, lines, circles, etc. we are working with things which really do have an existence, and really play their part, just as much as the 'real' things which are so much more evident to our imagination.

20. Higher Order Curves

20.1 One-to-two and two-to-two correspondences

So far we have worked with the basic relationship, the one-to-one. But it is possible to do other and more complex things. On a one-dimensional form (that is, a form composed of ∞^1 elements) such as a line or a conic of points, or a point or a conic of lines, one can establish a relationship between each element and a pair of mates in some given involution. It is most easily done the following way. We choose a conic on which to place our relationship but we know that the laws which emerge from our study will also apply to other forms, such as the line and the point — etc. We have conic α, point A on it and point B elsewhere on the plane (Figure 20.1). Now we know that each line through A will meet the conic in one other point only, but that each line through B will meet the conic in a pair of points in involution (Section 9.3).

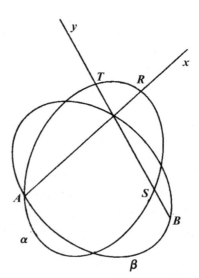

Figure 20.1

If we now put the lines of the pencils in A and B into one-to-one relation we shall get a correspondence between the single points and the involutary pairs of conic α. If x and y are corresponding lines in the pencils in A and B, then we say that point R, considered as a single point, corresponds with points S and T, considered as an involutary pair. Each of points S and T, considered singly, will, of course, correspond to quite other pairs.

If any point, considered singly, is found to be coincident with one of the points of its corresponding involutary pair, it is said to be an *incident member* of the relationship. We now have to ask ourselves the fundamental question: 'How many incident members will such a relationship produce?' This is similar to the question we asked ourselves in Section 5.4: 'How many self-corresponding members are there in an ordinary co-basal one-to-one projectivity?' It will be remembered that the answer was that there are two, and only two such; and we have seen that this answer has played a vital and far-reaching role in all that has followed.

The answer to our present question comes quite easily. We have already envisaged that we have a one-to-one projectivity between the lines of A and of B, and we know that the meets of corresponding lines will lie on some conic, β. It is clear that an incident member will be produced where the two conics meet, and we know that two conics always meet in just four points. However, the conic is bound to go through A and B, so we see that the meeting point at A is different from the other three. Conic β passes through A simply because A is the centre of one of the pencils engendering it. The line of B which passes through A will correspond to a line of A which will, in general, meet the conic α in some quite other point, and which will, of course, be tangent at A to conic β. Point A can only be an incident member if the two conics share this line as a common tangent; in this case they can only meet in two other points. In either case the answer is that there will be just three incident members.

If point A is not on conic α, but is some general point of the plane, the relationship becomes one in which one involutary pair is made to correspond to another involutary pair. In such a case it is clear, from what has gone before, that the number of incident members will always be four.

These facts lead the way to a number of important and beautiful constructions of higher order curves.

20.2 A construction of cubics

For instance, let us take a four-point system of conics and put any arbitrary point Q in the plane. Through Q there will be two tangents to each conic. If we mark in all the points of contact of all such tangents, on what curve will they lie?

Consider the pair of tangents to any one of the conics. The line joining their points of contact is the polar of Q with respect to that conic.
Now we know (Section 12.3) that all these polars must be concurrent.
Q lies on just one of the conics of the system, and through Q there will
be just one tangent to that conic. This tangent will touch just one other
conic, and the point of contact with that conic, R, is the point through
which all the polars of Q must pass. It is clear that we have a one-to-
one relationship between the polars of Q and the conics of the system.
Now let us take any arbitrary line, m, of the plane. The polars of Q will
make a range of points on it, and the conics will make an involution.
This relationship will make a correspondence between single points
and involutary pairs, and there will, therefore, be three incident members. These members are obviously the points in which our curve
meets the line. The curve is, therefore, a *cubic*.

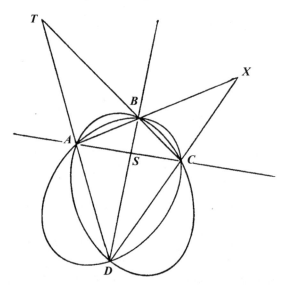

Figure 20.2

By examining this figure carefully, and visualizing to yourself the
various forms which the conics may assume, and especially the way in
which they will degenerate into the line-pairs *AB/CD*, *AC/BD* and
AD/BC you will soon convince yourself that the cubic must pass though
A, B, C and D, and also through the three points S, T and X, of the diagonal triangle (Figure 20.2). It must also pass through Q and R, the meeting place of all the polars. If we now take another point of the line QR
this will produce another cubic, but one which will also pass through A,
B, C, D, S, T, X, Q and R. By taking a selection of points along the line
QR we produce a pencil of cubics through nine common points.

Exercise 20a

One of the easiest and most beautiful ways of drawing such a system is to let our four-point system of conics be Apollonian circles and to meet them by any line we wish. This is the system pictured in Figure 19.3. A particularly pleasing effect is gained if our line passes through *Y'*, the midpoint between *S* and *T*. In such a case notice that the cubics degenerate twice — once into a circle paired with the line *Y'X*, and once into a hyperbola paired with the line at infinity. See also Figure 20.3.

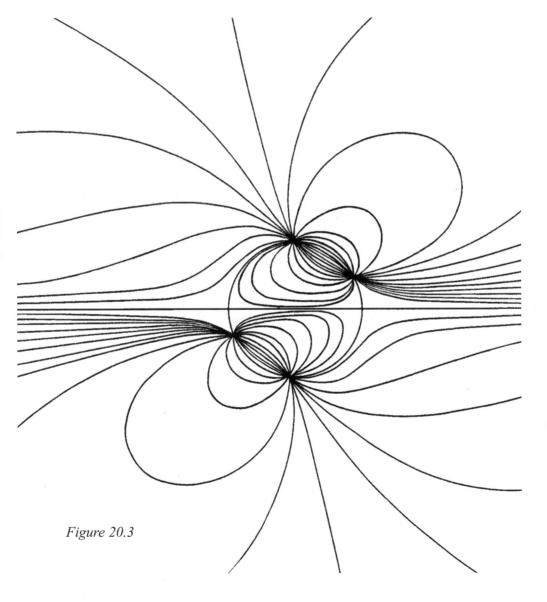

Figure 20.3

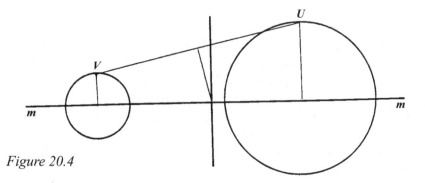

Figure 20.4

20.3 To find the common chord of two circles

We have already seen a method of finding the common chord of two real circles, in Section 14.3. This method is general, and will succeed in all cases; however, from Euclidean geometry we know that if the circles have real common tangents, their common chord will meet these common tangents at their midpoints between their points of contact with the circles. This is mentioned because in a drawing like that of finding the imaginary parts of Cassini curves (at the end of this chapter) the utilization of this fact can save quite a lot of time.

If the two circles are imaginary, draw their conjugate circles and join their centres by line *m* (Figure 16.4). Mark the ends of the radii which are perpendicular to *m*, *U* and *V*. We now have to find that point of *m* which is equidistant from *U* and *V*. This point must lie on the perpendicular bisector of the line *UV*. Where this bisector meets *m* is the point through which the common chord must pass, and a perpendicular to *m* through this point is the line we are seeking.

20.4 The curves of Cassini

We can now proceed to construct the real and imaginary points of a *lemniscate*. We start by constructing two sets of concentric circles, their radii increasing in geometric series — each set having the same constant ratio. Next we take the meeting points of any two of these circles. Now we move outwards one circle in one set and inwards one in the other, and take the new intersection points so produced. If we continue this process indefinitely what sort of curve will be produced?

Let us remember that concentric circles are a four-point system of conics (Section 16.4). Therefore, each of our sets of concentric circles will mark out its own involution along any chosen line. Now, two geometric series (growth measures) of equal constant ratios form a projective, one-to-one, relationship. Therefore, as we move from one circle to another along the geometric series we are establishing a one-to-one relationship between the circles of one set and of the other. Each circle puts an involutary pair on the line, and since the circles are in one-to-one correspondence, we have a one-to-one correspondence of involutary pairs on the line; that is, a two-to-two correspondence of points. Now we know that such a relationship will have four incident members and clearly these incident members will mark the points in which our curve will meet our chosen line. The curve we are constructing is, therefore, one of the fourth order.

By starting first with one pair of circles and then with another we are able to construct a whole set of curves belonging to one family. They are all of the fourth order, and are known as the *curves of Cassini*. The central curve of them all is the *Cassini lemniscate*.

Exercise 20b

Construct such a set of Cassini curves.

Exercise 20c

Construct just one of the Cassini curves, putting in all the common points of corresponding circles whether real or imaginary. Remember that we must visualize amongst them all the imaginary circles whose conjugate circles coincide with the real ones which we have drawn.

21. Path Curves

21.1 Collineation of the plane

We refer to Chapter 7, where we studied the growth measure. It will be remembered that there we set up a one-to-one transformation in which we transformed a line back into itself. We found that in such a case there are always just two invariant points — points which are self-corresponding within the transformation. If these are real it is a true growth measure, but if they are imaginary it becomes what we called a circling measure. Now we take any point of the line and watch how it moves as this transformation is applied again and again.

This growth measure is a truly archetypal process. If one of the invariant points is removed to infinity, it reveals itself as ordinary multiplication. The growth measure becomes a geometric series. But if the two invariant points are coincident the measure becomes what we called a step measure. This is the archetype of addition and it shows itself as this when the pair of coincident invariant points is at infinity.

Now we must ask ourselves what happens if we transform a whole plane into itself by a projective collineation. Will there be any invariant points, and if so, how many? And how many pairs of points can we choose at random in order that the transformation shall be completely determined?

Let us consider any two points, A and B, of the plane. And let the transformation be such that they transform, respectively, into A' and B'. Any line through A will transform into some line of A', and the pencil of lines in A will transform into a pencil in A', which will be, line to line, projective with the pencil in A. The meets of corresponding lines will form a conic which we will call the A-conic. Similarly, there will be a B-conic.

Now what can we tell immediately about these conics? Notice that line AB must clearly transform into line $A'B'$. Therefore the A-conic, besides passing through A and A', will pass through the meet of lines AB and $A'B'$. The B-conic will meet the A-conic here; and, of course, there will be three other common points of these two conics, real or imaginary.

At this point it is clear that the transformation is determined, for we can take any general point, *M* of the plane, and can construct its corresponding point *M'*, into which it will transform, as shown in Figure 21.1.

Now what is needed to determine the *A*-conic? It is three pairs of corresponding lines in the pencils at *A* and *A'*. One such pair we already have; *AB* corresponds to *A'B'*. Therefore, if we put down two further pairs of points, *C* transforming into *C'*, and *D* into *D'*, we have enough to determine the *A*-conic; and the same data will, of course, determine also the *B*-conic.

Thus we have the answer to the second part of our question. Any four points, *A*, *B*, *C* and *D*, can be put down, and can be made to transform into any other four points. *A'*, *B'*, *C'* and *D'*, and the transformation is completely determined.

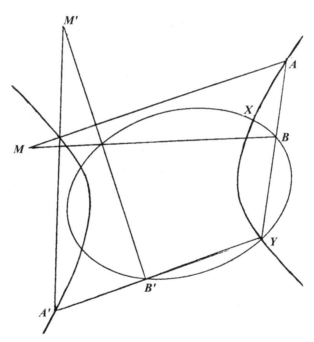

Figure 21.1

It follows from this that if we put down four invariant points, we have completely determined such a transformation without imparting any change to it at all. In such a case it is evident that no true change can take place; the transformation is the identity; no points move at all.

Thus we say at once that there cannot be more than three invariant points.

Now consider again Figure 21.1. Consider a point X, where the two conics meet. A moment's thought will show that such a point must be an invariant point. For before transformation, the point X can be considered as determined by the meet of rays AX and BX, while clearly after transformation it will be determined by the meet of the corresponding rays, $A'X$ and $B'X$. This is true of all the common points of the two conics except point Y, where the rays AY and BY are coincident. Y is a special point as far as this figure is concerned, in that this figure is not able to tell us where it will transform. We are thus left with three certain invariant points, and one doubtful. But we have already proved that we cannot have more than three invariant points, therefore, we may be sure that this fourth one is, in fact, not invariant.

Thus we have the complete answer. There will always be just three invariant points, and the movement of any fourth point being given the whole transformation is completely determined.

One finds that the invariant entities of a transformation must be self-dual, and this fact becomes of considerable importance later when we come to consider these transformations in a three-dimensional space. In the present instance it means that not only do we have three invariant points, but also three invariant lines. In fact, an invariant triangle.

Let us consider such a triangle ABC, with lines abc (Figure 21.2). Line a, considered as a whole entity, will be invariant. But the points of a will all move, each one into another point of a, except for the

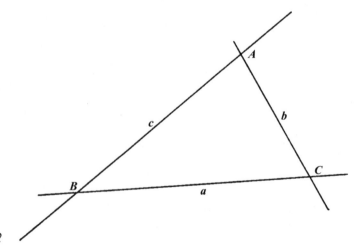

Figure 21.2

two invariant points *B* and *C*; these will be the fixed points of a growth measure along line *a*. Similarly for lines *b* and *c*. Similarly, point *A* is an invariant point, but the lines of *A* will all move, in growth measure between the invariant lines *b* and *c*. And the movement of these lines will be in direct perspective with the movement of the points along line *a*. We thus see how completely self-dual the whole transformation is.

21.2 Path curves

In order to construct such a transformation it is not necessary to construct the *A*- and *B*-conics. It is easier to do it this way. Put down any three points for an invariant triangle (Figure 21.3). Having done this we need to take only one more point, *M*, and move it at random to position *M'*; then the whole transformation is determined, and can be constructed.

Join *AM* and *AM'* to meet *a* in *P* and *Q*. Since *A* is invariant, and *M* transforms into *M'*, it follows that *AM* transforms into *AM'*, and therefore *P* into *Q*. Along line *a* we now have a growth measure determined, with invariant points *B* and *C*, with *P* moving to *Q*. In Section 7.2, we saw how to construct such a measure. Similarly, by joining *CM* and *CM'*, we can construct the growth measure of the transformation between *A* and *B* along line *c*.

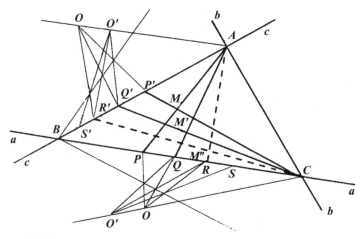

Figure 21.3

The same transformation which changes *AP* into *AQ*, will turn *AQ* into *AR*. Also this same transformation will change *CP'* into *CQ'*, and *CQ'* into *CR'*. Therefore, the transformation will take *M* into *M'*, and *M'* into the meet of *AR* and *CR'*, *M''*. By continuing in this way we find that point *M*, in successive stages of the transformation passes along a curve. This is called a *path curve.**

It is a wonderful thing. We have only to select an invariant triangle, and then move one point to a new position — at random. A complete transformation is now determined — and every other point of the plane starts, automatically, to move, each along one of a series of path curves which cover the entire plane.

By joining *A* and *C* to a whole series of points in the growth measures opposite to them we get a transformation grid, and by moving diagonally across the little quadrilaterals so formed we can draw a family of path curves.

By taking the harmonic mates, with respect to *B* and *C*, of *P*, *Q*, *R*, *S* etc. along the external portion of line *a*, we can complete the transformation movement along that line. And by doing the same along line *c* we can extend our grid to cover the whole plane.

Such a family of curves forms perhaps one of the most elementary and fundamental things in our world of thought. It is primeval in conception. It is simpler than circles or conics, and normally it includes neither, although in certain special cases it can do so. We have simply to postulate a one-to-one projective transformation — possibly the simplest and most primary mathematical process we can find. Now we apply the same transformation over and over again.

The result — inevitably — is a system of path curves. This picture — a general system of path curves — is so important that we illustrate it here (Figure 21.4).

Having made this figure it is well that we should consider it carefully, and ask ourselves what in it, apart from the actual invariant triangle, is invariant. And a moment's thought will show us the remarkable fact that the invariant element here really comprises the whole set of path curves itself. For consider, at any step of the transformation, each point moves on and takes the position of the next point on the curve, and simultaneously its place is taken by the preceding point of the curve; it is the same with all the tangent lines to the curves.

---

* The original German name is *W-Kurve*, which was misinterpreted as *Weg-Kurve* and then translated to *path curve*.

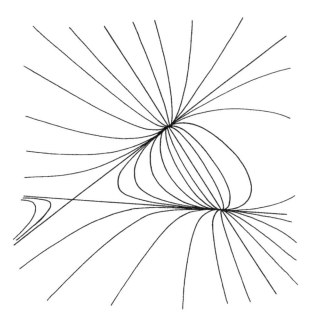

Figure 21.4

Apart from the three points and lines of the invariant triangle, itself, everything moves; but at the end of the move, nothing has changed! The whole set of curves is just as it was before. This is why the path curves are called the invariant curves of the transformation. No other curve we could conceive, in the whole plane, would have this quality; under the transformation it would change both its position and its shape.

This quality of the path curves can remind us of a similar quality which we find in the living organism. Any living organism lives just by a continual interchange of its substance with that of its environment. Something else in the organism, superior to the substance, transcends this continual change, and expresses itself in the invariance of the form. It takes substance, moulds it, uses it, and passes it on.

21.3 More about cross ratios

In Section 4.1, we learnt how to measure a cross ratio of four points in a given order. By taking the points in different orders there are, in fact, 24 different ways in which a cross ratio can be measured from any given set of four points. These 24 different ways give six different values — four ways to each value. If the value measured in one way is λ, then the other ways give values of

$$\frac{1}{\lambda}, \quad 1-\lambda, \quad \frac{1}{1-\lambda}, \quad \frac{\lambda-1}{\lambda}, \quad \frac{\lambda}{\lambda-1}$$

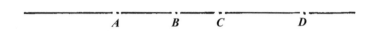

$$\begin{array}{cccc} A & B & C & D \end{array}$$

Figure 21.5

In that section we envisaged going from A to C, first via B and then via D (Figure 21.5):

$$(ABCD) = \frac{AB}{BC} : \frac{AD}{DC}$$

We could in fact, go from any point to any other, going first via one, and then via the other of the remaining two points. We could, for instance, go from A to B, first via C and then via D.

$$(ABCD) = \frac{AC}{CB} : \frac{AD}{DB}$$

But equally we could have gone from A to B, first via D and then via C. And both these cases could have been done, going from B to A. Thus, the journey between any two letters can be made in four ways. But we have six possible pairs of letters between which we can travel: A to B, A to C, A to D, B to C, B to D, and C to D. Four ways for each of these gives us 24 different ways of getting a cross ratio.

Figure 21.6

Now let us consider a growth measure, in which A moves to B, between the invariant points X and Y (Figure 21.6). We know that, in this projective process, the cross ratio of the four points $XABY$ remains constant through all the positions which A and B successively occupy during the stages of the transformation.

If we calculate our cross ratio by going from X to Y, first via B and then via A we shall get a certain number, and this number can be considered as distinctive of this particular growth measure. This cross-ratio would be written as

$$(XBYA) = \frac{XB}{BY} : \frac{XA}{AY}$$

and would be calculated as

$$\frac{XB}{BY} \times \frac{AY}{XA}$$

If we now project Y to infinity, we know that the successive positions of A and B in the growth measure will form a geometric series, and we shall find that the constant ratio or multiplier of this series is identical with the cross ratio which we have just measured. Had we measured our cross ratio going from X to Y, first via A and then via B, we should have got a result which would be the reciprocal of the first. In other words. instead of moving up the geometric series, multiplying by, say, x, we should be moving *down* it, multiplying by $1/x$.

In the following work we shall consider that our cross ratios are calculated in this way, and we shall put an arrow showing the direction of the movement.*

---

* Note that this cross ratio ($XBYA$) equals that of Section 7.3: ($YAXB$).

So
$$(XBYA) = \frac{XB}{BY} : \frac{XA}{AY}$$

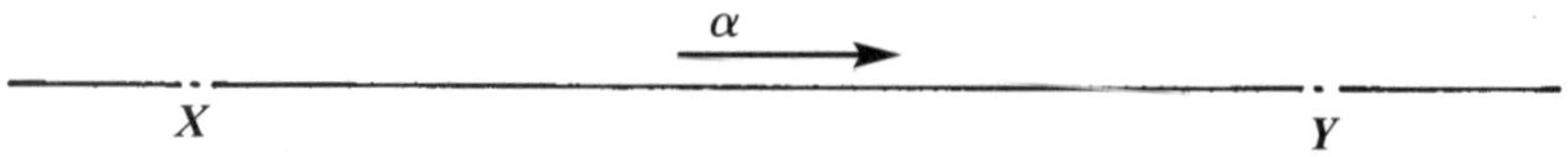

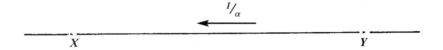

Figure 21.7

We shall indicate a growth measure between two invariant points like Figure 21.7; meaning that if we move along the measure in the sense of the arrow we are, in essence, multiplying by the factor α, but if we were to move in the opposite direction we would have to draw it like Figure 21.8.

These last two figures are really equivalent to one another; that is, the same series of points (or numbers) is implied, but we are moving along it in different directions in the two cases.

Figure 21.8

21.4 Cross ratios round the invariant triangle

We now come to an important theorem. We have a triangle ABC, and two points of the plane, M and N. We project these points on to a, b and c, from A, B and C respectively. We will call the cross ratios along these lines α, β and γ respectively (see Figure 21.9), going counter clockwise around the triangle). In such a case it will always be found, no matter how M and N are placed, that the product of these cross ratios equals unity.

$$\alpha\beta\gamma = 1$$

In order to prove this we have only to add a few more letters, and one line, to Figure 21.9, as shown in Figure 21.10.

Projecting the cross ratio on line a from point A, we can see that

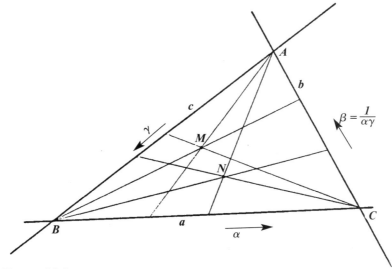

Figure 21.9

it becomes equivalent to a movement from M to T along line SC, and this, projecting from B onto b becomes equivalent to a movement from U to V along line b. Similarly the cross ratio along line c, projected from point C becomes equivalent to a movement from T to N along the line through A; and this, again projected from point B onto line b, becomes equivalent to a movement from V to W along that line. Thus, the product of these two cross ratios is exactly equivalent

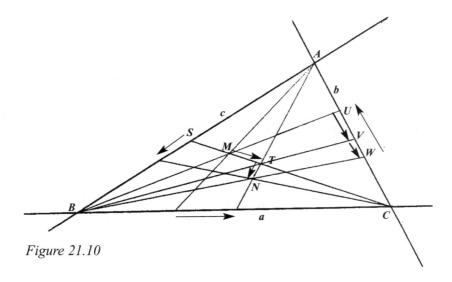

Figure 21.10

to the cross ratio along line *b*, except that it goes in the reverse direction to our standard anti-clockwise direction of turning round the triangle — that is, it is the reciprocal of that cross ratio.

Exercise 21a

Write down the above proof in detail.

This reversal of direction along one side of the triangle is always to be found in path curves of the plane case; we shall always find that two of the potencies go one way round, and the third goes the other way. Thus, we could redraw our figure to look like Figure 21.11, and this is the true representation as far as the path curves are concerned.

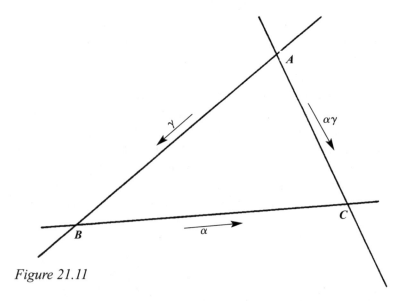

Figure 21.11

In fact, we can say that the multiplier of the growth measure which goes in the opposite direction from the other two, will always be the product of the multipliers of the other two.

21.5 Some algebra of path curves

Notice now in the drawing of the path curves in Figure 21.4, that all the curves pass through two of the invariant points and avoid the third, and similarly they are all tangent to two of the invariant lines, and none of them is tangent to the third. The line which they all avoid is the one whose growth measure goes opposite to the others, and the point of avoidance is the point opposite to that line.

If we work in *homogeneous co-ordinates*, and take the invariant triangle as the triangle of reference,* the equation of a family of path curves is

$$x^a \, y^b \, z^c = k$$

where a, b and c are the logarithms of α, β and γ. As k varies, we pass from one curve of the family to another. (Since $\alpha\beta\gamma = 1$ it follows that $a + b + c = 0$, and this renders the above equation homogeneous.)

It follows from this that if any two of the cross ratios are given, say α and β, then the third, γ, is already determined, and with it, of course, the whole form of the family of curves. Such a result was to be expected from our drawings of Section 21.2; we only had to construct two growth measures in order to construct all the curves.

Thus we have only to know the relationship between two of the growth measures, say α and β, to know the whole shape of the curves. The relationship in question is not one of simple ratio, but of what one might call *exponentional ratio*. We do not want to know what ratio α is of β but what *root* it is of β. It is the ratio of the exponents that counts.

For instance, if $\alpha=16$ and $\beta=4$, a certain shape of curves will result, and the points during transformation will move round their curves with very big steps. If $\alpha=9$ and $\beta=3$, the same shape of curves will result, but the points will move round them with *smaller* steps. We can imagine α being reduced until it is only just more than unity; we can still find a value for β, which would be $\sqrt{\alpha}$. We should then have the points moving in minute steps around their curves. We arrive at the true plastic path curve when we let α, in the limit, approach as near to unity as it can, without actually becoming unity, and still, of course, keep β as the square root of that. The truly smooth movement along a path curve comes from what we may call an infinitesimal transformation.

It will be noticed that the *ratio* of the exponents, a, b and c, in this equation above will determine the shape of the curves, and this agrees with what we ordinarily know of homogeneous equations.

---

* In the Euclidian plane one uses Cartesian coordinates, which require two axes of reference. In the projective plane one uses homogeneous coordinates, which require three axes.

We now have to examine what sort of curves these are. If we take the general case — that is, we put down the points *M* and *N* at random in the plane of our invariant triangle — there is a high degree of probability that the values of *a*, *b* and *c* will be irrational to one another; that is, no one of them can be expressed as a perfect fraction of either of the others. In such a case we cannot speak of the resultant curves having any order; they are *transcendent curves*. Where *a*, *b* and *c* are rational to one another; that is, can be expressed as whole numbers, then we can put them in as exponents in the equation and read the order of the curves in the usual way.

We reproduce in Figure 21.12 just two of the curves from the path curves drawing of Figure 21.2. We see point *M* becoming point *N*, and from there, through the stages of the transformation, approaching nearer and nearer to the invariant point *C*. Now we ask the question: After the curve passes through *C* which way does it go? Does it curve round by the heavy arrow? Or does it pass through an inflexion and follow the light arrow? Or does it pass through a cusp and follow the dotted arrow?

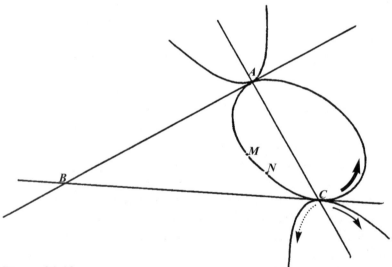

Figure 21.12

Firstly, we must examine to what an extent this is a legitimate question to ask. If I ask where the point *M* goes to in the successive stages of the transformation after it has passed through *C*, the question is meaningless, because the whole nature of the path curve formation ensures that the point will *never reach* C. The successive stages of its movement will grow smaller and smaller as they approach C, and only after an infinite number of stages can we imagine the point reaching C. Thus the answer to this question is that an infinite number of points are continually streaming into C along all branches of the curve which pass through C, just as an infinite number are streaming outwards from A.

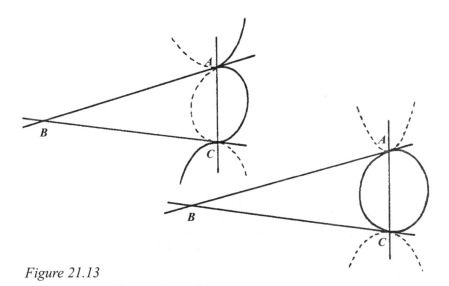

Figure 21.13

But a curve need not be considered just as a stream of points, but as an entity in itself — one expressed, for instance, by its particular equation. And regarding the curve on which M and N lie as an entity, we are entitled to ask where this curve goes as it passes through C. The strange answer to this question is that, in the figure as it stands, there is nothing to tell us which way the curve goes. The curves are always so shaped that they can either be flexed and cusped as in the left of Figure 21.13 or they can pass through as in the right example.

If the curves are transcendent; that is, if a, b and c are irrational to one another, this question must be forever unanswered; in fact, there is no answer. We can only think of points streaming from all curves *out* from one invariant point and *into* the other. The curve as an entity has not yet completely come to earth as it were.

If, however, a, b and c are rational to one another, they can always be expressed as whole numbers, and we suppose that these have always been reduced to their simplest form. If the largest of these (disregarding signs) is odd — the curves will be flexed and cusped but if the largest is even they will go round as in the right side of Figure 21.13.

We can most easily understand the way things go by considering a concrete case. Let us suppose that $a=1$, $b=2$ and $c=-3$. We then have the equation,

$$x\, y^2 z^{-3} = k$$

and the curve will be flexed and cusped. The actual shape of the curve will be given by the ratio between any two of the values, say $a{:}b = 1{:}2$.

But suppose we had the equation

$$x^{101}\, y^{201}\, z^{-302} = k$$

This will give us a curve of almost the same shape as the previous one, as $a{:}b = 101{:}201$ is very nearly the same ratio as $1{:}2$. But it would now be a curve like that in the right side of Figure 21.13, without cusp or flex. By using very large numbers we could have got the ratio $a{:}b$ as close to $1{:}2$ as we wished, while still keeping the index of z an even number.

Thus we can say that as we let the ratio $a{:}b$ grow gradually greater than its present value of $1{:}2$ we would pass rapidly through an infinite number of shapes of curve, alternately of the forms seen in Figure 21.13. And in between each we would pass by an infinitude of transcendent possibilities, where the numbers are irrational to one another.*

* We should keep in mind that x^a is defined only if a is integer or if $x \geq 0$. So in the end we should admit that
— $x^a y^b z^c = k$ generally represents a curve inside the 'positive' part of the plane
— $|x|^a \lfloor y|^b |z|^c = k$ represents four path curves
— if a, b and c are integers $x^a y^b z^c = k$ may represent two or four path curves.

21.6 A special case

Let us suppose that the constant cross ratios along sides a and c of the invariant triangle are the same; that is, that $\alpha = \gamma$ (Figure 21.14). Then the two growth measures form projective ranges, and the pencils of lines in A and C are projective pencils. Therefore, the curves must be conic sections. From Section 21.4 it is obvious that the points of line b move in a growth measure with cross ratio $\alpha\gamma = \alpha^2$.

But suppose, instead of making M transform into N, we had made it transform into N'. This would have given the same growth measures exactly, except that the one along c would have moved in the opposite direction. In such a case we notice that as the lines of the pencil in A approach the line AC, the lines in the pencil in C approach the line CA. In the limit the line AC of the A-pencil corresponds with the line CA of the C-pencil. In other words, we have two projective pencils whose common line is self-corresponding. We know that this means that the two pencils are in direct perspective (Section 5.2), and meets of corresponding lines lie on a line. Thus in this particular case, if we follow the little quadrangles along the other diagonals from those which gave us the conics, we shall get straight lines, radiating from B.

By proceeding thus we are drawing the path curves of a quite separate transformation from the one which produced the conics. We

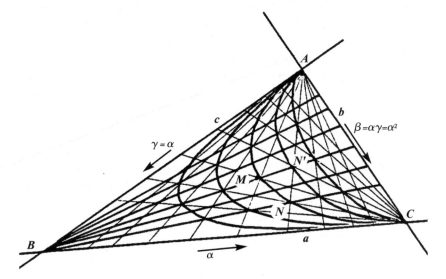

Figure 21.14

thus see that, in a special case, path curves may be straight lines radiating from one of the invariant points. In such a case we see that all the points of the plane move in growth measures along lines radiating from *B*, except the points of line *b* itself; all these are now self-corresponding or invariant. All the lines of *B* have become invariant, considered as lines, and the invariant triangle, as such, has becomed engulfed in the general invariance of the pencil in *B*. All that remain distinctively of the invariant triangle are *B* and *b*.

But we have already met such a case. In Section 3.3 we learnt about the transformation which we called homology. This now turns out to be nothing but a special case of path curves.

Had we tried moving the other way across the quadrangles in any other case of path curves than the conic one, we should have found ourselves moving along a new set of path curves, given by some new transformation. It is only in the case of the conics that this new set of curves degenerates into straight lines.

Notice that when we reversed the direction along *c* in our figure we should have had to write $1/\alpha$ against line *c* and

$$\alpha \times 1/\alpha = 1$$

against line *b*. Unity here means that the transformation along line *b* is equivalent to multiplying by 1; nothing moves; we have identity.

If we now ignore the construction lines in Figure 21.14 we have a grid made of curves and lines. We know that the lines, in their spacing, make a growth measure between lines *a* and *c*, and it is clear that the curves meet these lines in growth measures between *B* and *b*. Thus if we now move diagonal-wise across the lozenges of the grid we shall produce some curves of yet some other transformation, possible within that invariant triangle. It can be shown that these are path curves too. By moving from one line to another, but missing out every other curve we should find yet a further set of path curves. In fact, there are any number of sets of path curves which can be found by varying the way in which we move across this grid. This principle is used widely later on.

Exercise 21b

Draw a set of conic path curves. This can be easily done by making a growth measure along, say, line *a*, and then projecting these points from any point of *b* (conveniently the point at infinity) on to *c*. This ensures that the cross ratios along the two sides are the same. Draw the conics and the straight lines associated with them, and then, moving diagonally across the grid so formed, draw another set of path curves, belonging to some other transformation of that invariant triangle.

21.7 Special features

Let us refer once again to the right-hand drawing of Figure 21.13. These curves have neither flex nor cusp, but that is not to say that they have no singularities. In the general case, a special thing happens to the curve both at *A* and at *C*. At one of these points the curve, for an infinitesimal stretch of its length, inclines to become straight, that is, to lie along the invariant line c; while at the other end, just for an infinitestinal moment, the tangent moving round the curve tends to turn within point *C*. The first of these singularities we shall call a *flattening* and the second a *sharpening* and they are clearly dual to one another. Only in the case of the conic do these features altogether disappear.*

There are two special configurations of the invariant triangle which give interesting and very important figures. Firstly, we can let *B* move out to infinity. Lines *a* and *c* then become parallel and the growth measures along them become, of course, ordinary geometric series. The resulting figure is so important that we picture it here (Figure 21.15).

We let line *b* be at right angles to *a* and *c*, and we let the constant ratio along *c* be considerably greater than that along *a*. We choose the movements so that as *a* moves inward, *c* moves outwards. We can see clearly the flattening and the sharpening at *A* and *C* respectively.

Had the constant ratios been the same along the two lines these curves would have been ellipses within the lines and hyperbolas

---

* These flattenings and sharpenings are not singularities in a mathematical sense, except in degenerate cases.

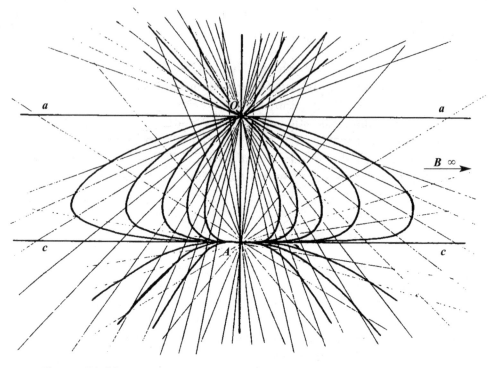

Figure 21.15

without. These special cases are seen in inanimate nature (planetary orbits, the movements of projectiles etc.) but in the living nature we are constantly reminded of the general case, with its tendency to flatten at one end and sharpen at the other (eggs, pine cones, etc.).

A second configuration of the invariant triangle which is of interest is when *B* remains on the centre of the page and *A* and *C* remove to infinity. We will let them be at right angles on the line at infinity. The growth measures along *c* and *a* again become geometric series, and the curves look like Figure 21.16. If the constant ratios along the two lines are equal, all the curves become rectangular hyperbolae.

Exercise 66

Redraw Figures 21.15 and 21.16, without having any pair of lines at right angles.

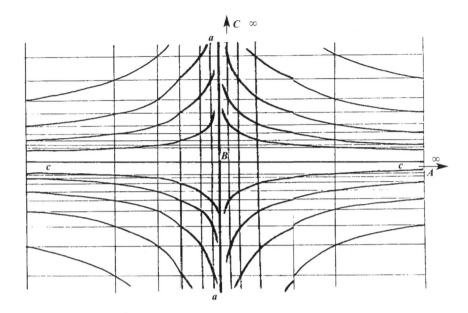

Figure 21.16

21.8 The imaginary invariant triangle

In Section 16.2, we saw how to make an imaginary triangle that is, a triangle having two of its points and lines conjugate imaginary, and one of each real. To begin with, for our triangle, we will have the line at infinity as the real line, and the imaginary points, I and J, as invariant points on it. The third, and real, point of the triangle will be point O, in the middle of our page. We let I be the clockwise rotation on the infinite line, and J the anti-clockwise. Then the clockwise rotation of the right-angled involution in O will be line $j=OI$ and the anti-clockwise will be line $i=OJ$. This is the celebrated triangle OIJ (Figure 21.17).

Now within this triangle we wish to put the grid, of curves and lines, which we had in Section 21.6, Figure 21.14. Notice that in that figure, B corresponds to the point we are now calling O; A and C have become I and J; and line b has become the line at infinity of our new figure. The path curves of that figure were conics through A and C; these will become circles in our new figure, as they go through I and J. The conics of Section 21.6 meet the radial lines

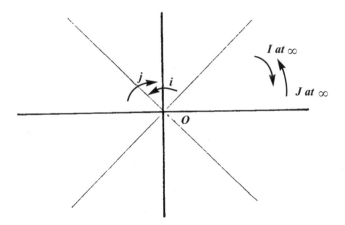

Figure 21.17

through *B* in growth measures between *B* and *b*. Thus we must now draw our circles in growth measures between *O* and the line at infinity; that is, with their radii in geometric series. We thus see that concentric circles are nothing more than a set of path curves with respect to this particular invariant triangle.

Now we must remember that circling measure with respect to the invariant points *I* and *J* manifests itself as equiangular steps. In Figure 21.14 the radial lines through *B* met line *b* in a series of points which were in growth measure between *A* and *C*. So now the radial lines through *O* must meet the line at infinity in circling measure between the points *I* and *J*. That is to say that we must draw our radial lines with equal angles through *O*.

Having done this we have constructed our grid, exactly equivalent to that in Figure 21.14. And now we can draw another set of path curves, belonging to the same invariant triangle, by just going diagonal-wise across the little lozenges of the grid, see Figure 21.18.

But by doing this we move in equiangular measure around *O* while we move outwards in geometric series. This is to construct nothing else but a *logarithmic spiral*. So we see that a set of logarithmic spirals is, in fact, a set of path curves with respect to the imaginary invariant triangle, *OIJ*.

21.9 The equi-angular quality

Now look again at Figure 21.14. Imagine the tangent to the curve, at *M*. This tangent will meet the line *AC* in some point, and the line *BM* meets line *AC* in some other point. In one step of the transformation the points in which this tangent and the line *BM* meet *AC* will each move one step along line *AC*. From the laws of the one-to-one transformation it follows that the cross ratio of these points with *A* and *C* will remain unchanged. But we know that an unchanging cross ratio manifests itself in the circling measure of *I* and *J* as an equiangular movement. Now we take any point on a logarithmic spiral and join it to the centre, and we consider the angle which this line makes with the tangent at the point. It means now that if we let the point move around the spiral this angle will always remain constant. This is one of the laws of the logarithmic spiral which we can find by metrical methods, but which is obvious straight away as soon as we see the curve as a path curve with respect to the triangle *OIJ*.

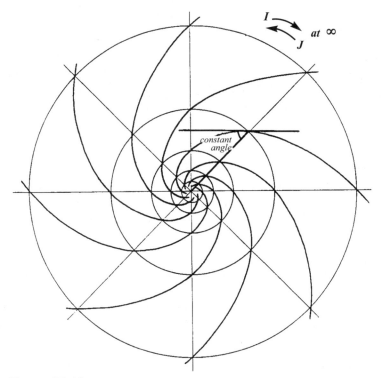

Figure 21.18

Exercise 21d

Construct the set of path curves with respect to some other imaginary triangle, not *OIJ*. In Section 17.2 we saw how to construct a set of concentric circles in perspective; that is, a set of conics all passing through two given conjugate imaginary points and all tangent to two given imaginary lines which are in perspective with those points. These conics which must be constructed in growth measure between the common centre and the absolute line, *z*, form a set of path curves with respect to that particular imaginary triangle composed of the given imaginary points and lines. Now join the centre point to the points of the circling measure which you have already constructed along the line carrying the pair of imaginaries. By moving across the diagonals of the little lozenges so formed you will be able to draw the spiral curves with respect to this triangle. This is an exercise well worth doing, and gives a very beautiful set of curves with a pronounced rhythmic quality.

Figures 21.4 and 21.18 appear so different, but they are, in fact, projectively exactly equivalent. Notice how, in Figure 21.14, when we go diagonally across the lozenges of the grid, we have a curve which goes through *B*, and either through one of *A* or *C*, avoiding the other. In Figure 21.18 we see the curves going through (or towards) point *O*, and always curving in either one or the other of the two possible directions. The curves we have drawn, pass through *I* but not through *J*.

In Section 21.6 we saw how the points on the curve crowd closer and closer together as they approach the invariant points, and do not reach them until an infinite number of steps of the transformation have been taken. This manifests itself in the *OIJ* figure in the fact that the curve winds forever inwards towards the invariant point *O* but cannot be said to reach it until an infinite number of turns have been made.

We saw in Figure 21.15 how alive and organic the path curves can look, and now here again we find that in another expression of the same process we come on one of the fundamental forms of the living organic world.

Exercise 21e

Projective spirals can be constructed in a slightly different way.

Start with an elliptic movement (circling measure) on the line z and label the points $A_0 \ldots A_{17}$. Take any point in the plane, not on the line, and label it O. On OA_0 take an arbitrary point X_0. Connect this with, say, A_4. The meeting point of this line with OA_1 is the next point of our spiral, X_1. Repeat this process, connecting X_1 with A_5 to get X_2 on OA_2, etc.

Exercise 21f

Instead of connecting X_0 with A_4, you could connect it with some other point, for instance A_6. Repeat the construction. Note that the shape of the spiral has changed. You could end up with the conics of Section 17.2. How?

21.10 Degenerations

A point that moves on a spiral has a double motion: it rotates about the centre and at the same time it moves away from the centre O, towards the line z, which we can call expansion. The curvature of the spiral depends on the ratio of the speeds of these movements: a big rotation and a small expansion gives a big curvature, a small rotation and a big expansion gives a small curvature. In the limiting cases, for instance, if the rotation becomes zero, the spiral degenerates to a straight line, and our path curve system becomes the homology of Section 3.3. If at the other hand the expansion vanishes we end up with the concentric conics of Section 17.2 (or 16.4 for that matter).

21.11 The intermediate case

We remember that in Figure 15.5 (page 167), we learned of the relationship of a line to a conic, when that line moves parallel to itself. As the line approaches the real part of the circle the amplitude of the imaginary points of the circle on that line grows narrower and narrower, and in the moment when this amplitude becomes nil the points have coalesced into one, and have at the same time become real; the line is now a tangent to the circle with its real double point of contact. The line has only to move a fraction of a space a little farther and the two real points become distinct, with a growth measure implied by the circle between them. We thus see that when the two invariant points of a growth measure coincide the resulting form constitutes an intermediate stage between the growth measure with two distinct real points, and the circling measure with two distinct imaginary points; and we know that this intermediate stage is the one which we have called *step measure*.

It is important to bear in mind that although the growth and circling measures look so very different from one another they are, in fact, projectively quite equivalent. The fact that the controlling elements have become imaginary instead of real in the case of the circling measure makes no vital difference to the essential mathematical properties of the process. But the momentary, intermediate, stage is something quite different in nature — less elementary and more complex. This in spite of the fact that the step measure in its superficial appearance bears considerable resemblance to the growth measure.

We see the metamorphosis in Figure 21.19. At the top we have the growth measure, between invariant points X and Y. Notice the direction for the movement, notice also how both X and Y act as blocks to the movement, and that the movement is always in opposite direction on the two sides of an invariant point.

Now if in imagination we let X and Y draw nearer together, so that the movement in the inner section between X and Y has less and less space, we can come to the moment when these two points coincide. The double invariant point still acts as a block, but the movement is in the same direction on both sides of the point.

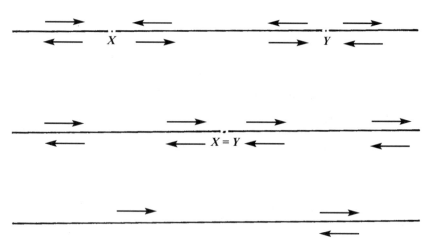

Figure 21.19

Now remove this double block, or rather let it float into the imaginary, and the two movements can go smoothly through, in circling measure. In fact, with Von Staudt, we can say that the movements, with all the possible circling transformations which they imply, are the invariant points X and Y.

A similar metamorphosis can be taken with the invariant triangle of the plane path curves (Figure 21.10). These five pictures show the metamorphosis. In the first two the points B and C are real, but com-

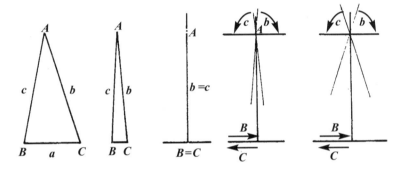

Figure 21.20

ing close together. Then they have merged, and we have the intermediate stage; in the fourth they have become imaginary but are still very close to one another; that is, their amplitude is very small, and in the last they have gone much further off into the imaginary world.

Exercise 21g

It is a very enlightening thing to draw in full the path curves for the middle three pictures. The first and last of these we already know how to do, although they will now look very extreme in form, owing to the shape we have chosen for the invariant triangle.

21.12 Exponential curves

The remainder of this chapter is devoted to considering the middle
picture of Figure 21.20. Notice that this triangle is self-dual like all
the others; it has two points: a single point A and a double point
$B = C$. It also has two lines: a single line a, and a double line $b = c$.
Along line a the points will move in step measure, and along line b
they will move in growth measure between A and B. Similarly, the
lines of A will move in step measure, with line b as invariant line, and
the lines of point B will move in growth measure between invariant
lines a and b. At the moment these various measures are not yet
determined, but we have now only to say that point M moves, in the
transformation, to point N, and the whole thing is fixed (Figure
21.21). Join AM and AN and where these lines meet line a, we have
two points of our step measure; and we can easily now construct the
rest of this measure. Then join the point B to M and to N, and these
two lines are lines in the growth measure of lines between a and b as
invariant lines. This whole measure can then be constructed, by using
a Steiner circle (Section 6.16) or by any other means we wish. We

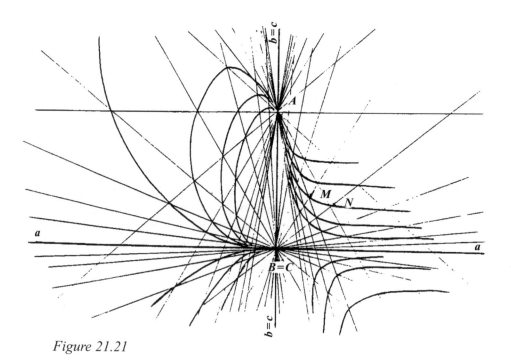

Figure 21.21

then have our grid, and a series of path curves can be quickly drawn. On the other side of line *b* the step measure may be drawn symmetrically with respect to point *B*, and the growth measure harmonically with respect to lines *a* and *b*.

You will notice when you have drawn this that the two parts of the figure, to left and to right of line *b* are quite unsymmetrical, and that already the curves look as though they want to start spiralling, but that their movement is blocked by the presence of the double line *b* and the double point *B*. When the right hand picture of Figure 21.20 is filled out with path curves the general shape of the curves is very much the same over most of the plane, but the blocks have been removed and a real spiralling movement is there.

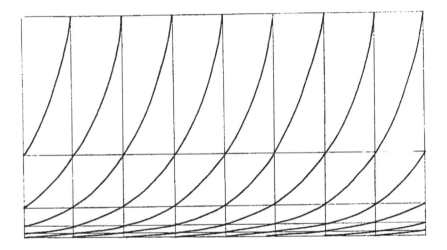

Figure 21.22

Now what sort of curves have we in Figure 21.21? We have curves formed by the playing against one another of a growth measure and a step measure. These take on a more familiar form if we let the double line, *b*, move to infinity, *a* remaining the same (Figure 21.22).* Point *A* moves to infinity also, of course, but we put it at right angles to *a*. The measure of lines in double point *B* will now be parallel lines in growth measure between line *a* and the line at infinity *b*. In other

---

*   Note that line *a* is not shown. It is a little bit below the figure.

words we have parallel lines in geometric series. The lines through
A, however will clearly meet line a in the equal steps of an arithmetic
series. What we have is the *exponential curve* of the equation

$$y = e^{(cx+k)}$$

where the constant ratio of the geometric series of the parallel lines is
e^c. By varying k we move from one curve to another of the system. If
we now vary c, we shall get a new set of curves produced by moving
at a different speed in the horizontal direction from that in the vertical
— say one step horizontally to two vertically, or two horizontally to
one vertically, etc. Having settled on a new value of c we can move
through all the curves of the new transformation by again varying k.

We thus see that these exponential curves also reveal themselves as
being essentially path curves; and the figure where the line b is finite
and on the page reveals interesting side branches to these curves which
we would hardly have expected from the ordinary drawing of Figure
21.22.

Exercise 21h

A slightly different way to construct these curves is as follows.
— Project M and N onto a and construct the step measure as
    before.
— Let AM and a meet in S, and AM and BN in T (indented) M and
    T determine a growth measure between A and S, which can be
    constructed by the method of Section 7.2.
— Join A to the points on a and B to those on AS to get the grid.
— Draw the curves.

21.13 Summary

So far we have met the following path curve-systems:
— the *triangular system*, see Section 21.2
— the *exponential (or logarithmic) curves* of 21.12*
— the *spirals* of 21.8
— the *concentric conics* of 17.2 (see also 21.10)
— the *homology* of 3.3 (see also 21.10)

---

* Since exponential and logarithmic curves have exactly the same shape (each is
  the reflection of the other), both names are used.

There are two more types. If we move the centre of the homology onto its axis, we get the *elation* (Section 3.3). Its path curves are lines through the centre.

If we move the centre of a system of concentric conics onto its absolute line, we end up with a system of *osculating* or *contangential conics* which is treated in the next chapter.

22. Osculating Conics

22.1 Playing with homology

We have seen (Section 21.6) that if the cross ratios along two of the sides of the invariant triangle of a one-to-one transformation are equal, then the resulting path curves are conics, sharing those two sides as common tangents. This is so only if the cross ratios move in the same sense round the triangle. If one of them reverses its direction — so that they both move away from their common point, the curves all degenerate into straight lines and we have, in fact, a very special case of our transformation. All the points stream outwards in growth measure from B towards the line AC; all the lines of B become self-corresponding and so do all the points of AC; the invariant triangle loses its identity, and we have left a transformation which is, in fact, identical with the *homology* which we described in Section 3.3.

Now let us suppose in an ordinary homology, that we put a conic on our plane, to pass through O, the centre of homology, and to meet the invariant line of the homology in S and T (Figure 22.1). Now we take any point M of the conic and let it transform into M', any point chosen along the line OM. The transformation is now fixed and we know that the conic will transform into another conic. Since O, S and T are self-corresponding points we know that the new conic must share these with the original one, and since all lines of O are self corresponding,

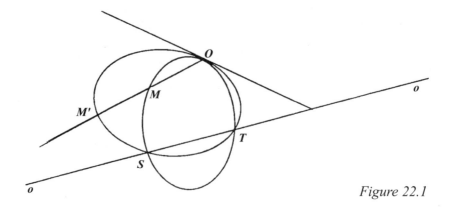

Figure 22.1

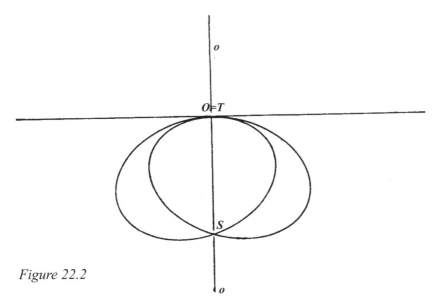

Figure 22.2

the line through O which was tangent to the first conic will also be tangent to the second. In other words, the four points which these two conics share in common will be S, T and a double point at O.

Now let the invariant line, o, turn about S, until it passes through O (Figure 22.2). The other point in which it meets the conic will slide round until T is coincident with O. We then have an *elation* (see Section 3.3), and we can construct the new conic in exactly the same way as before. Now the new conic must, of course, still have four of its points in common with the original one, and from the whole way the construction has proceeded it is clear that three of these are coincident at O, and the fourth one is at S. Such a pair of conics are said to have *three-point contact* at O.

Now we know that any four points determine a single infinitude of conics (Section 12.1) so that the triple point at O must determine at least ∞^2 conics. Among these infinitudes of conics there will always be one that is a circle; this is clear from our Euclidean knowledge of the circle — any three points determine just one circle. This is the *circle of curvature*, and its radius is the usual way of expressing how great the *curvature* of the curve is at this point.

22.2 Four-point contact

If we next turn the line OS about O, we can let S slide round until it is coincident with the triple point OOT. In other words, we have an

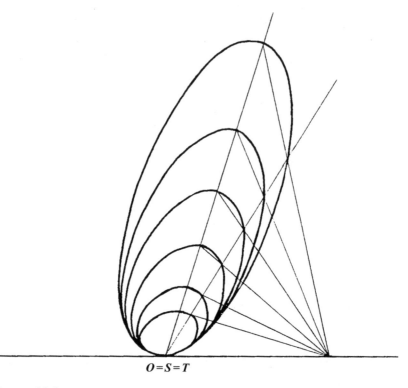

$$O = S = T$$

Figure 22.3

elation and we put our conic down to pass through the invariant point *and* to be tangent to the invariant line (Figure 22.3).

We then transform our conic into a series of conics, and we shall have curves which touch one another in four coincident points, the quadruple point *OOST*. We know that five points completely determine just one single conic, therefore, it is clear that they cannot have more than four points in common, either separately or coincidentally, without becoming identical curves all the way round their length. Thus these conics, with four-point contact at *O*, touch one another with the greatest intimacy, at that point, which it is possible for them to have.

Two curves which touch one another with this greatest degree of possible intimacy are said to *osculate* one another.* This is a family of *osculating conics*. It is just as truly a four-point family (pencil) as

---

* In modern mathematics the 3-point contact of the previous section is already called osculation. Higher order contact is sometimes called *hyper-osculation.*

the one which we pictured in Figure 12.2 (page 138), where the four points were distinct; and all the geometrical laws of such a family, which were described there, apply here also. For instance, any line will meet these conics in pairs of points in involution (Desargues' Conic Theorem, Section 12.2).

Notice the beautiful way in which these conics gradually swing the direction of their major axes, as they grow larger.

22.3 Some properties of osculating conics

If we draw any two lines through O these will meet each conic in a pair of points which will determine a chord of that conic; and all such chords, determined by one pair of lines through O, will be concurrent at some point of the invariant line o. This follows immediately from the method of construction used to draw the figure. If we allow the two lines through O to come together we get the property that if we draw any one line through O, the tangents to the curves where this line meets them will all be concurrent at some point of the invariant line.

Notice that the conics of this figure stretch out towards infinity, and, if we care to draw them all, past it; it, therefore, consists of many ellipses and hyperbolas, and one parabola, but nowhere in it can we find a circle. However, if we had put the original conic down so that point O came at the end of either the major or the minor axis, then we would have found just one circle in the family. We may say that, in general, four infinitesimally close points on a conic are not *concyclic*, but that at the ends of the major and minor axes, they are.

Notice that if we have curves with triple contact they always change sides with one another, but that curves with four-point contact do not. If we were to try to picture, in a very crude way, what is happening in the immediate (that is, the infinitesimal) neighbourhood of the point, it would look something like Figure 22.4, although, of course, the inflexions are not present.

3-point contact   4-point contact

Figure 22.4

22.4 The Segre Type 3

We must now consider path curves with the ultimate degeneration of the invariant triangle: that is to say when the three invariant points have melted into one.

There is a convenient way of representing the state of the invariant triangle. If it is non-degenerate — that is, the three points are distinct — we say that it is Type 1 1 1. If two of the points have coincided, leaving one distinct, we call it Type 2 1 (two-one); and if all three points are coincident we call it Type 3. Such numbers are called the *Segre characteristic* of the transformation in question. In Section 21.12, we dealt with Type 2 1.

It will be remembered that we let points B and C draw closer and closer together (Figure 21.20), but that even when they were infinitesimally close they still determined a quite definite line, BC, between them. Thus, in the limit, we were justified in still drawing a quite definite line BC 'joining' the coincident points $B = C$, and in putting along it, a step measure.

Carrying on from Figure 21.20, we now let A move gradually down towards the double point $B = C$. Even when it is infinitesimally close, but still just distinct, there is still a definite line joining the double point to A. If we now let A coincide with $B=C$ we shall have a triple point, containing two lines — a single line BC and a double line $AB=AC$.

Now we must move very warily. In Section 21.1, it was mentioned that the invariant elements of a one-to-one transformation must always form a self-dual figure. But this one is not, being formed of one triple point, and two lines — one single and one double.

Let us now try to construct such a transformation as the one which we have envisaged. Since each of the two invariant lines contains only one invariant point we must put a step measure along each, marking the direction of movement by arrows. Now with any point M we ask where it will move next. An obvious method of construction presents itself. Through M draw the two lines a and b. Now one step in the transformation will change a into b and b into a new line, c, joining the next pair of points in the step measures. The meet of b and c will mark the new position of M.

Now all step measures are projective one with another since they are each projective with a set of equally-spaced points, so along with our two invariant lines we have projective ranges of points, and it is clear

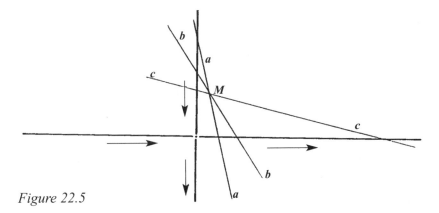

Figure 22.5

that the triple invariant point is a self- corresponding member of the two ranges. Therefore, they are in direct perspective, and joins of corresponding points will be concurrent. Therefore, the point ab is identical with the point bc (Figure 22.5). We see that our 'transformation' has frozen into a sort of immobility. In fact, no such transformation can be said to exist. We must try some other way if we are to attain Type 3.

This time we will let our triangle collapse as in Figure 22.6. A comes down towards the double point $B = C$ while B and C are still just distinct. The lines AB and AC then spread outwards so that in the last moment when all three points are about to merge, all three lines do likewise. We then have an invariant triangle formed of one triple point containing one triple line. This is self-dual, and we can proceed in safety.

We put down our invariant point O (the triple point $A = B = C$) and invariant line, o. The lines of O will be in step measure from o, and the points of o will be in step measure from O. (The projective way to draw a pencil of lines in step measure is to dualize Figure 8.5 (page 112); the lazy way is to join the centre of your pencil to equally spaced points on some chosen line!) We have still one more free choice. A

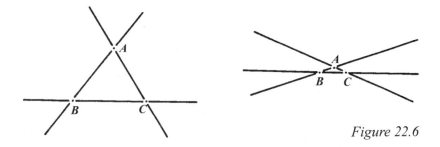

Figure 22.6

point *M*, on one of the lines of the step measure through *O*, can move
to any point *M'* on the next line of the step measure. Now the transformation is completely determined.

For the sake of convenience we will let the line *MM'* meet *o* in one
of the points of the step measure transformation which we have actually marked, *A* (Figure 22.7). Now under transformation *M* transforms
into *M'* and *A* into *B*. Therefore, the new point *M"* will lie on the line
M'B, and will obviously be where this line meets the next line of the
step measure through *O*. Proceeding in this way we can easily construct the rest of the path curve, and we shall find that it is tangent to
o at *O*. Further curves in the family may be easily obtained by letting
BM' meet *OM*, *CM"* meet *OM'*, etc. We will name these two points of
the second curve, *N* and *N'*.

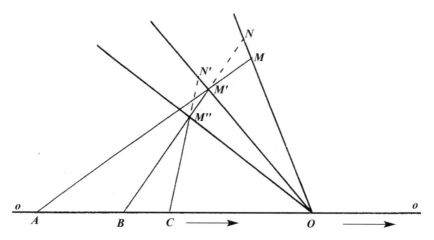

Figure 22.7

Now we must enquire what sort of curves these are. Let us consider
the tangent to the first curve at *M*. It will meet *o* at some point, say *X*.
This tangent will transform into the tangent at *M'*, while *X* will transform into the next point of the step measure along *o*, say *Y*. It follows
that as the chord of the curve through *O* moves in step measure (*OM*,
OM', *OM"*, etc) the tangents at the end of this chord meet a tangent of
the curve (*o*) in step measure also. All step measures are projective, so
the range *XY*... along a tangent of the curve is projective with the pencil *OM*, *OM'*... etc. through a point of the curve. Now this is a distinctive property of the conic section, and we will not be surprised to learn
that the curves of Type 3 are always conics, tangent to *o* at *O*.

Now let us consider Figure 22.7 simply as an abstract figure, and not as a drawing of path curves. We will let NN' meet line o at K. We will keep all points fixed except M'', which will be moveable along line NB, N' which will be moveable along OM', and K which will consequently have to move along o. As M'' moves along BN, N' will move along OM' in direct perspective with it from centre C. As N' moves, K will move in direct perspective with it from centre N. Therefore, K will move along o projectively with M'' along BN.

When M'' is at M', K will be at B; when M'' is at N, K will be at C; when M'' is at B, K will be at O. But B, M'', M' and N are harmonic, since they lie on 3 successive step measure lines and the double line of the measure. Therefore, when M'' lies on its original position, being the harmonic conjugate of N with respect to B and M', K must lie on the harmonic conjugate of C with respect to O and B. But this we know is A (since A, B, C, and O are three successive points in a step measure and the double point of the measure).

We have now proved that with such a construction as we have made for the two path curves $M...$ and $N...$ the lines MM' and NN' are bound to be concurrent on the invariant line o. But this is precisely the construction we used in Figure 22.3, in order to transform a conic by elation into another having four-point contact with it. We, therefore, see what the path curves of Type 3 are; they are a family of conics having four-point contact at the invariant point and with the invariant line for common tangent. In fact, of course, they have four-line contact on line o.

Exercise 22a

Most interesting figures can be made by repeating this construction putting O at infinity and letting line o be the line at infinity; and then for another figure putting O at infinity and letting line o run across the middle of the page.

23. Fundamental Notions for Geometry in Space

23.1 Duality and dimensions

We must refer first to Chapter 2, where we dealt with the *principle of duality*. A re-reading of that chapter would be a good introduction to our present studies. We remember that the fundamental entities of space are three — points, lines, and planes, and that the first and last of these are completely dual to one another; whatever we may say about points, we may also say, in a dualistic way, about planes. We see straight away that the *line* holds a unique position in our thought, quite different from points and planes. For instance, any three points, whatever, have a plane in common, and any three planes, whatever, have a point in common; there are no exceptions at all. Any two points, whatever, have a line in common; so do any two planes, whatever — again there are no exceptions. But when we come to consider two lines the position is immediately different. Any two arbitrary lines may meet — or they may not. In the first case, when they meet, they will have both a point and a plane in common; they cannot have the one without the other. In the second place, they have neither in common and we say they are *skew*.

Now let us refer to Section 16.1. There we introduced the idea of orders of magnitude; it would be a good thing to revise the first two paragraphs before going further here. We saw there that we can say that a line contains ∞^1 points, whereas a plane contains ∞^2. We are very near here to the fundamental truths about *dimensions*. We can say that the plane is a two-dimensional form since it contains a two-fold infinity of points — also of lines. We come to a deeper idea or 'dimensions' than that given by just the three right-angled directions of Euclid.

In Chapter 2, we saw that a plane figure can be polarized in two quite different ways. We can polarize it within the plane, simply changing point for line and line for point. We then get a new plane figure. But we could also polarize it within the context of three-dimensional space — all points of our figure will become planes, all the lines will remain as lines, and the whole lot, instead of lying in one plane, will pass through one point. Our plane figure will change into a point-centred figure. A triangle polarized within the plane will change into another, or the same, triangle — it is a self-dual figure. But polarized within three dimensions it changes into a trihedron — three lines and their common planes, all lying in (passing through) one point. A pointwise conic, with its attendant tangent lines, polarized within the plane becomes a linewise conic, with its points of contact of the tangents.

A conic polarized in three-dimensional space becomes a cone, of lines and tangent planes, all passing through one base point.

It is interesting to notice what happens to the ideas of *order* and *class* when we polarize a figure. Remember that order and class are dual concepts. Thus when we polarize the order of a plane curve we shall come to the class of a cone. The order is the number of points in which a line meets the curve; the class is the number of tangent planes of the cone which are held by any arbitrary line of the base point (Figure 23.1).

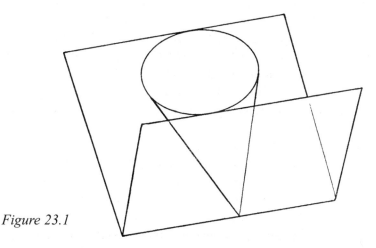

Figure 23.1

On the other hand, the class of a plane curve is the number of tangent lines which are held by any point of the plane; thus the order of a cone is the number of lines of the cone which lie in any arbitrary plane of the base point (Figure 23.2).

We thus see that, from this aspect, the point is a truly two-dimensional form. It has just as much richness of possibility within it as has the plane. We often see books of Plane Projective Geometry; we could equally

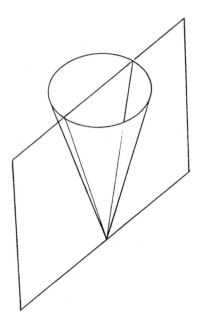

Figure 23.2

well, but not equally easily, have a Point Geometry. Every figure and every theorem in the plane would have its dual counterpart within the point; the logic and the reasoning would be identical, but it would be much harder to visualize. The plane is an extensive entity; the point is an intensive one. Plane geometry is an extensive relationship of largely intensive elements, but point geometry would be an intensive relationship of extensive elements. Our consciousness is so formed that it comparatively easily comes to an understanding of extensive forms, but the intensive ones come much harder to it. Yet the reasoning is just as easy in the one case as in the other.

Notice that the conic, in the plane, is a one-dimensional form — it has a single infinitude of points and tangent lines. Similarly, the cone, within the point, is a one-dimensional form containing a single infinitude of lines and tangent planes. Thus we say that, considered as a form of lines and planes, a cone is a one-dimensional figure; but considered as a form of points it is two-dimensional; it has infinity lines, and each line has infinity points, all on the cone; therefore it has ∞^2 points. If we are to visualize the cone as a one-dimensional form within a two-dimensional point geometry, we would have to consider all its lines and planes *in the way in which they actually pass through the point*. All the points of those lines and planes which lie outside the base point would have to be completely ignored; they would not form part of the figure at all. This way of visualizing would be necessary for understanding an intensive geometry. If we consider a cone two-dimensionally; that is, as a surface of points, what is its dual? For this one would have to start with a linewise conic in a plane, and then try to visualize all the planes passing through each tangent line. The resulting form, in as far as it can be held in the imagination at all, does not strike one as being probably very fundamental or significant. But it is just such a form which one is considering when one visualizes a cone as a surface of points.

Rudolf Steiner once said 'We must learn to see the extensive intensively, and the intensive extensively' and in these two types of geometry we see the way towards such thinking.

Notice that a sphere is a three-dimensional form, but that the *surface* of a sphere, containing ∞^2 points and ∞^2 tangent planes, is a two-dimensional form.

Now we must ask ourselves how many points there are altogether in space. Let us consider any line. It contains ∞^1 planes. It is evident to our imagination that these planes will fill all space. Each plane contains ∞^2 points. This leads us to the idea that there are ∞^3 points in space. By dualizing the above conception one easily arrives at the result that there are ∞^3 planes in space.

When we come to lines, however, the case is different. How many lines are there in space? Consider any plane. Every line of space will meet it in just one point. Therefore, if we consider all the lines which meet this plane we shall have considered them all. The plane contains ∞^2 points, and through each of these points pass ∞^2 lines of space. We thus see that there are ∞^4 lines in space. Again it is the line which has a unique position, and this fact leads to many very important consequences. From this also we begin to get an idea of what is meant when we read in some book of the 'four dimensional space of lines.'

(In dealing with all the above, when we say ∞^2, ∞^3, etc. it is most important to be constantly reminding ourselves that we are not dealing with quantities, but rather with concepts which may be regarded more as orders, or dimensions, of magnitude, as explained in Section 16.1).

23.2 Curves and surfaces

In plane geometry, after one has studied figures made up of points and lines, one moves on to plastic forms. There is only one kind of plastic form in the plane; that is the *curve*, although, of course, every curve is made up of points and lines, and can be considered both pointwise and linewise. In space there are two kinds of plastic form, the *surface* and the *twisted curve*. The first is a two-dimensional form, being made of ∞^2 points and ∞^3 tangent planes. Commonly seen examples are the sphere and the spheroid (rugby ball).

As long as a curve stays in a plane it is known as a *plane curve*, but as soon as it leaves one plane and journeys out through space it is called a *twisted curve*, or a *space curve*.

23.3 The anatomy of a plane curve

We must now ask ourselves how a point moves in a plane. The answer is that it must move, momentarily, along a line. It is not possible to imagine it moving in any other way. Let us suppose that at some particular moment a point is moving along a given line (that is, the line shows the direction of its movement at that moment), but that at the same time the line itself is moving around the point, so that the very next moment the point is already moving in a different direction. Then the point would be moving along a curve and the line which shows its momentary direction of movement is the tangent line of the curve at that point.

But next we must ask how a line moves in a plane. The answer is the dual of the above case. The line must always, momentarily, move in (or turn about) a point. Even if the line moves parallel to itself it is still moving in a point — the point at infinity of the line. If then at any particular moment we have a line moving in a given point, but at the same time that point is itself moving along the line, then the line is enveloping a curve, and the point about which it is moving is its point of contact with the curve.

We are now in a position to appreciate what we might call the anatomy of a curve. A curve is caused by a point moving along a line, while at the same time that line is moving in that point. Or, dually, it is caused by a line moving in a point while at the same time that point is moving along the line. The two statements express exactly the same thing, seen from the two dual aspects. In the first case the line along which the point moves is the tangent to the curve at that point, and in the second the point around which the tangent moves is its point of contact with the curve.

Since the line is turning about the point we may say that that point is the common point of two infinitesimally close positions of the line; and since the point is travelling along the line we may say that the line is the common line of two infinitesimally close positions of the point. Thus we return to the definitions for a tangent which we had in Section 2.8.

23.4 The anatomy of a space curve

When we come to three-dimensional space things are a little more complicated. We must still say that a point always moves, momentarily, along a line, but we must now add that a plane also moves, momentarily, around a line. (If a plane moves parallel to itself it is moving around its line at infinity.) However, when we come to consider the movement of a line, we again find that the position is quite a different one. A line *may* be moving in such a way that it is always moving, momentarily, around, or in, a point. If it is doing so, let us consider two infinitesimally close positions of the line. They have a point in common — the point around which the line is moving at that moment. But we know from our fundamental proposition that two lines which have a point in common also have a plane in common. It follows, therefore, that these two infinitesimally close positions of the line lie in a common plane, and, therefore, that the line is moving, at that moment, in that plane. Thus we may say that if a line is moving in such a way that it is always turning around some point, it is also always turning in some plane. In such a case the movement of the line envelops a twisted curve, and a plane in which it is at any moment moving is called an *osculating plane* of the curve.

We may now enunciate the 'anatomy' of a twisted curve. It is formed by a point moving along a line, while that line is moving in the point, and is also moving in some plane; but simultaneously this plane is turning within the line and the point. Or dually, a twisted curve is formed by a plane which is moving within a line, while that line is moving within the plane, and also moving in some point; but simultaneously the point is moving along the line and within the plane. These two things make up exactly the same form, and are in fact, only two aspects of the same thing. The plane is the osculating plane of the curve at that point.

23.5 The ruled surface

However, seeing that two lines need not meet at all, it is clear that it would be possible for a line to be moving in such a way that even two infinitesimally close positions of the line still have no common point.

They could be that much closer together and still be skew. The line is then said to be *moving skew to itself*. Such a movement of a line does not envelop a curve, but traces out a whole plastic surface in space. We thus have the possibility of surfaces which are everywhere curved, but through every point of which pass one or more straight lines which lie, throughout their length, completely in the surface. Such a surface is called a *ruled surface*, and the lines tying in it are called *rulers* or *generators* of the surface.

23.6 Order and class

We already know what is meant by the order and the class of a plane curve. We now have to understand the equivalent ideas for three dimensional forms. The *order* of a surface is given by the number of points, real or imaginary, in which it is met by any line, and the *class* of a surface is the number of tangent planes of that surface which pass through any line. If we now imagine our surface to be, let us say, of third order, and suppose that it is cut by some plane, we see that every line of that plane will meet the surface, and therefore, also the section of that surface by that plane, in three points. It is clear that the surface will therefore meet every plane of space in a third order curve; and this is a general rule for surfaces of any order.

Let us think out the dual of this. We have a third class surface, and we confront it by any point of space. This point will contain a cone of tangent planes of the surface. But every line of this point will contain three tangent planes of the surface; therefore, we may say that every point of space will envelop the surface with a cone of third class.

The *order* of a twisted curve is the number of points in which that curve meets every plane of space. Dually the *class* of that curve is given by the number of osculating planes of the curve which pass through any arbitrary point of space.

In future, whenever we speak of two points, lines or planes which are infinitesimally close together we shall describe them as being *neighbouring*.

The single infinitude of planes which are contained in any line will be called a *pencil of planes*. The ∞^2 planes which are contained in any point will be called a *sheaf of planes*. The ∞^2 lines which are contained in any point will be called a *star of lines*.

24. The Quadric

The ideas described in this chapter are absolutely fundamental; they are also exceedingly difficult to visualize, or illustrate. It is strongly urged that the reader should make for himself a model such as is described in Section 24.5, and have it by him for constant reference while reading, or re-reading, it. The model is best made with two squares of acrylic or other transparent sheeting, for the top and bottom planes, held in place by four wooden or metal rods, one at each corner. The relevant points, equi-angularly spaced round the two circles can be drilled with a very fine drill, and then threaded with coloured threads.

24.1 Cross ratios

Let us consider 4 planes, α, β, γ and δ, all lying in one base-line m (Figure 24.1). Now let us cut these planes by two lines, x and x', which share a common plane, π. Line x will meet the 4 planes in 4 points, A, B, C and D, while x' will meet them in the 4 points A', B', C' and D'.

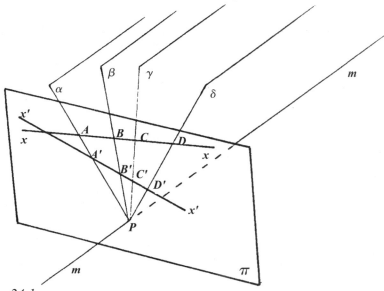

Figure 24.1

We let line m meet plane π in point P, and it is then clear that the 4 planes α, β, γ and δ will meet π in a pencil of 4 lines centred in P. It is also quite clear to the imagination that these lines will be the lines AA', BB', CC' and DD'. From this it becomes clear that the cross ratio of $ABCD$ equals that of $A'B'C'D'$.

Now let us imagine the four planes cut by two skew lines, x and y. If we take any other line, z, which meets both x and y, we will see from the above, that z meets the four planes in four points which have the same cross ratio as both the ranges on x and y. Therefore, the cross ratio along x will be equal to that along any skew line y. From this it is clear that the four planes will meet any line whatever in a cross ratio which is equal to that in which they meet any given line.

This is the cross-ratio property of a pencil of four planes, and it is the exact spatial equivalent of the cross-ratio property of a plane pencil of four lines.* It is of fundamental importance and from it we can see that a pencil of four planes possesses a pure number as its cross ratio — the cross ratio in which it is met by any arbitrary line of space.

24.2 Three skew lines

Let us consider three skew lines in space, a, b and c. How many lines can we have which will meet all three?

First let us take any point M of line b. M and a will determine a unique plane α. M and c will determine a plane γ. Planes α and γ will have just one line in common, m. This line will, of course lie in both α and γ and it, therefore, must meet both a and c. Since M lies in both α and γ it must lie on their common line, and will, therefore, lie in m. Thus by picking out the one point M of b we have found just one line, m, which meets both a, b and c. But there are ∞^1 different positions along line b which M could have had; therefore, there are ∞^1 different lines to be found, which will meet a, b and c.

We have reached an answer to our original question quite easily, but we must go on from here. There are ∞^2 points in a plane. In Section 6.1 we found a process which picked out for us a single infinitude according to a particular law, and we then found that this single infinitude of points made up a definite plastic entity. (We took the common points of corresponding lines in two projective pencils, and found that they lay on a conic.) Here we shall find that a similar thing occurs. The

---

* It is also the space dual of four points on a line, see Section 4.1.

process in this case is exceedingly simple: we take any three skew lines, and we find that there are ∞^1 lines which meet all three. Our three skew lines have picked out for us, from the ∞^4 lines which are possible in space, a single infinitude, and we shall find that this single infinitude of lines makes a definite plastic form.

Firstly, let us take two neighbouring positions of line m and let us assume that they have a point, and, therefore, also a plane, in common. Clearly both lines a and b must lie in that plane. If this be the case, then a and b, being coplanar, must meet. But we have already stipulated that they are skew. Therefore, it is clear that the two neighbouring positions of m can have no common point or plane. In other words, as M moves along line b, m moves through space, skew to itself. The plastic form with which we are dealing, therefore, is not a space curve, but is a surface — as described in Section 23.5.

24.3 The quadric surface

We now have to decide what sort of a surface it is. Notice that if we have four positions of point M — M_1, M_2, M_3 and M_4 — these will have a definite cross ratio and they will transfer this cross ratio to both the sets of planes α_1 ... α_4 and γ_1 ... γ_4. From this it is clear that the pencils of planes in lines a and c are projective; they are in one-to-one relation. Now we cut the whole form by any arbitrary plane of space, π, and we note what sort of section it makes in this plane. Lines a and c will appear as points, and the projective pencils of planes in them, as projective pencils of lines in those points. Lines m, being common lines of corresponding planes, will appear as common points of corresponding lines in these pencils, and these points will, by Section 6.1, lie on a conic. We therefore see that our infinitude of lines m, must lie on a form which meets every general plane of space in a conic. Our form must, therefore, be a surface of second order. It is called the *quadric surface*, and in the realm of surfaces it is the exact equivalent of the conic in the realm of plane curves.

It is clear that this surface is a completely plastic one, curved in every part, yet is crossed by an infinitude of straight lines, each of which lies for the whole of its length completely in the surface. Such a line is called a *generator* of the surface, and a set of infinity generators is called a *regulus*.

The picture so far is of a quadric surface covered by an infinitude of generators, each of which lies wholly in the surface, and each of

which is skew to each of the others. Meeting each of these generators are the three skew lines with which we started, *a*, *b* and *c*. These
are clearly also generators of the surface, since each obviously lies
wholly in the surface. If we now take any three lines of the set of
infinity generators we discovered to lie in the surface, we shall again
have three skew lines from which we can start, say, *x*, *y* and *z*.
Taking all the lines which meet these three we have a new regulus,
which will clearly include *a*, *b* and *c* within it. In fact, it transpires
that this new regulus lies on the same quadric surface as the first
one.

In Section 24.5 (Figure 24.3) we give a method by which a model
of such a surface can be made, and a rather diagrammatic picture of
one. It will be useful to bear this picture in mind while reading all
that follows this. In our picture, we have drawn only the one set. The
other set would cross this one symmetrically in the other direction.)

Our picture of the quadric now is this — a plastic surface, entirely
woven over by two sets of generators, so that through every point of
the surface run just two generators. Each generator meets every generator of the other set, but is skew to every one of its own set. Moreover,
every plane meets the quadric in a conic.

Now let us take any four generators of the first set; they have been
formed by the meets of four planes of line *a* and four corresponding
planes of line *c*. The cross ratio of these two sets of four planes will, of
course, be equal (they are projective pencils of planes) and it will be
equal to the cross ratio of the four points *M* in which they meet line *b*.
But every generator of the second (*abc*) set will meet each of these four
generators, and always in the points in which it meets the four planes
of *a* and *c*. It follows, therefore, that these four generators will meet
every generator of the other set in four points which have the same
cross ratio. This is a fundamental law of the quadric surface, and it is
very important. It means that any four generators of one set can be said
to have a cross ratio. It is the cross ratio in which they meet every generator of the other set.

This is an important spatial property which can be stated without
reference to the quadric. Take any three skew lines; take any four lines
which meet all three. These four will then meet each of the three in an
equal cross ratio.

24.4 Tangent planes

Next let us consider any two generators, one of each set. By what has gone before we know that they will have a common point and a common plane. Their common plane is a *tangent plane* to the surface, and this plane touches the surface in their common point. Thus we can say that through any point of the surface there run two generators, and their common plane is the tangent plane at that point. If we now let the plane through our generator slowly turn around that generator, its point of contact with the surface will move along the generator. But we know that every four positions of a generator of the other set has a cross ratio which is that of the four planes of the generator in which they lie, and that they impart this cross ratio to every line as they meet it. Therefore, the cross ratio of any four points of contact of the line is equal to that of their four tangent planes. As the point of contact moves along a generator, its tangent plane turns round that generator, *projectively with it*. Such a projective relationship of points and planes all belonging to one line we call a *parabolic strip*, and it is of great importance in the study of certain spatial transformations later. It appears here as one of the most fundamental laws of this most fundamental surface — that a forward movement along a line is associated projectively with a turning movement around that line.

Following from all this there is one fact which we need to hold as vividly as possible in our imagination. It is not easy to illustrate in a figure, but needs really to be appreciated with a string model in front of one. Let us consider any generator of one set, and any plane of this generator; this plane will meet the quadric in a generator of the other set. These two generators, being co-planar, must have a common point, which agrees with what we have already stated, that each generator of one set meets every generator of the other. This plane therefore must be a tangent plane to the surface, touching it at the common point of the two generators. As this plane turns about the original generator, it holds within it, in turn, each of the generators of the other set, until when it has made a half turn, it has covered in this way the whole of the surface of the quadric. In the course of this, the point of contact with the surface will have moved along the whole length of the generator, forming the parabolic strip which is described above.

Next let us consider any two planes of space. They will meet the surface in conics. The common line of the two planes will meet the surface in two points only (second order surface); therefore, it follows that these two conics will have common points, X and Y, on this line. Now from what has gone before it is clear that if we take any four generators of one set they will have a definite cross ratio and they will meet each conic in four points which will have this cross ratio on that conic. Thus as we move round the successive generators of either set we should be marking out projective ranges along the two conics, and the common points of those conics, X and Y, would be the double points of the projectivity.

24.5 The string model

This gives us a new way of generating a quadric surface. Take any two conics, in different planes, which meet on the common line of their planes (Figure 24.2). Take projective growth measures (that is, ones having the same constant cross ratio) along the two conics, between X and Y, the common points of the conics, as double points. Take any pair of points, one on one conic and one on the other, and connect them. Next connect successive pairs of points round the growth measures. The lines so chosen will be generators of a quadric surface, and they will form a growth measure of generators on that surface. (Notice that the double lines of this

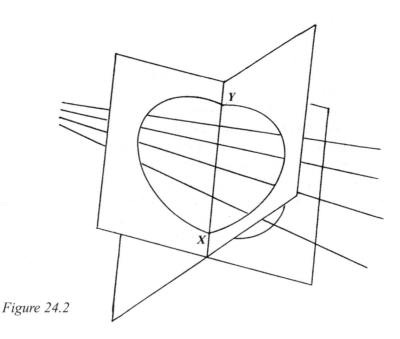

Figure 24.2

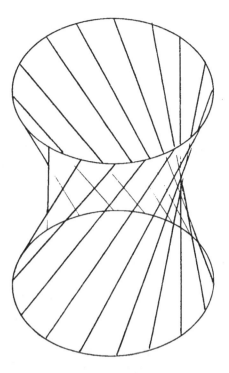

Figure 24.3

growth measure are not given by this method, but they are certainly implied, and could quite easily be found by further means).

There is nothing so far in our thinking which stipulates that these conics must meet in real points. The meeting planes might be so situated that the conics in which they meet the surface do not meet in real points. The position, however, is projectively not altered; they will meet in a pair of imaginary points which will lie on the common line of the two planes. Our growth measures between X and Y now become circling measures between the imaginary points in which the conics meet the line. The easiest way to construct this is to make the circling measure between these imaginary points along *the line* and then to project this on to the conics from the poles of that line with respect to the conics. Those who have completed Exercise 15b (page 169), will understand why this point must be chosen as the centre of perspective for this projection. This leads to a most important method of generating a quadric surface. Let the two planes be parallel; their common line will be their line at infinity (Figure 24.3). We can take circles for our conics, as these meet in the points I and J at infinity. Round these circles we must now put points in circling measure between I and J as double points; that is, they must be equally spaced round the two conics. Now we join any point of the top conic to a point

of the bottom one, say 50° round, and then continue joining points like
this, equi-angularly round the conics. This gives us our set of generators.
To obtain generators of the other set we join each point of the top plane
to one on the bottom which is 50° round in the other direction. It is not
necessary for the conics to be circles; they may both be ellipses, but in
this case they must be ellipses which have common points on the line at
infinity. That is to say that they must have the same *shape* (ratio of length
to breadth) and they must be similarly oriented; that is, their major axes
must be parallel. The exercise of constructing the circling measure of
points on these ellipses, with respect to the imaginary pair in which these
ellipses meet the line at infinity is left to the reader.

24.6 The linear congruence

We have seen that we have to consider that there are ∞^4 lines in all
space, and that if we have some method by which we single out a
single infinitude of these, this single infinitude will have some definite plastic form. In the method just considered we arrived at the regulus of lines, all lying on a quadric surface.

It is possible, however, to find methods which choose out ∞^3 lines
of the ∞^4 that are possible in space. The form arrived at then is called
a *complex*, and it is a highly complicated organism of lines. It would
not be possible for anyone to picture all that is involved in such a form,
all at once, but it has to be studied, and pictured, bit by bit, in one
aspect after another of its many-sided richness.

It is also possible to have methods which will pick out for us ∞^2 of
the ∞^4 possible lines of space. Such a form is called a *congruence*. It is
also exceedingly rich and many-sided.

Here is the simplest way in which we may come to an idea of a congruence. Let us imagine any two skew lines, a and a', and ask ourselves how many lines there are which meet both of them. Let us take
any point M of a. M can be joined to each of ∞^1 points of a', and there
are therefore ∞^1 lines through M which meet a'. But there are ∞^1 positions which M can assume on line a. Put any two skew lines before
one; consider all the lines which meet them both, and one has a *linear
congruence*. It is as simple as that!

Notice that if we have two skew lines, they pick out a congruence
(∞^2); but that if we introduce another limiting element such as a third
skew line, then they pick out a regulus (∞^1). The word *linear* used here
does not mean that it is a congruence made of lines — all congruences,

in this sense, are that. But imagine to yourself that you are viewing these two skew lines from some definite point of space. They would *appear* to meet at just one point. A moment's consideration of this fact shows you that through any point of space there passes just one line of such a congruence and for this reason the congruence is called *linear*; that is, a congruence of the first order, just as a linear equation is one of the first order. Other higher order congruences we are not concerned with at the moment. The congruence which we have just considered, that is all the lines which meet both of the skew lines a and a', will be referred to as the congruence aa'.

Now on our two skew lines, a and a', we will place growth measures, having double points X and Y on a, and X' and Y' on a'; and we will arrange that these growth measures have equal cross ratios. We join XX' and YY', and we take any one other pair of points, one of each growth measure, and we join them, say AA'. Now we have three skew lines, XX', AA', and YY', and these we know will determine a regulus on a quadric. We know that all the generators of the AA' set will meet both a and a' in projective ranges. But the growth measures which we have put down on a and a', being of equal cross ratio are already projective. If therefore we now proceed to join pairs of points in succession along the growth measures, from AA', we shall, in fact, be generating the quadric in question. Our generators will lie upon it in a growth measure of lines, with XX' and YY' as double lines.

We have here an alternative way of generating a quadric. It is formed of the joins of corresponding points in two projective ranges on two skew lines. Note that this is not really a new method of getting the quadric; it is already implied in what we have done up till now, but it is important to see it from this new aspect.

It is not necessary to use the growth measure for getting our projectivity. We could just as easily, and often will, use circling measures. We remember that in this case the equal cross ratio property shows itself in the equal-cycle property: if we have a 12-cycle on the one line it must be a 12-cycle on the other, etc. (See Section 13.3). Just as well we might have formed our projective ranges in any other way open to us, but we shall find that the use of growth and circling measures has special usefulness and convenience for us in our further studies.

We must take note that the particular cross ratio chosen for our growth measure, or the particular number chosen for our cyclic measure, has no effect whatever upon the shape of the quadric which will be produced. This is determined entirely by the choice of double points and of the first pair AA' which we join up. (In the circling measure the choice of the amplitude and centre would correspond to the choice of the double points in the growth measure.) These give us our three lines which determine the quadric; the choice of cross ratio determines the 'speed' with which we can imagine the lines of one set to move in the other.

Now we notice that this quadric is made up entirely from the lines of the congruence aa' and we can next ask the question, 'How many such quadrics would it be possible to find within the lines of the congruence?'

On the line a there are ∞^1 positions for Y. There are therefore ∞^2 choices for our two double points. For each of these choices there are ∞^1 different choices for our point A. Therefore, there are ∞^3 different ways of choosing X, Y and A. Clearly for each choice of X, Y and A there are ∞^3 ways of choosing X', Y' and A'. This gives us, at first sight, ∞^6 different quadrics. But we must be very careful! If we take any one of these quadrics we find that it contains many other trios of lines which could have determined it. We have just found that the ways of choosing three elements out of ∞^1 number ∞^3. Therefore, out of our ∞^6 choices, ∞^3 have gone to each separate quadric. The number of quadrics possible therefore is ∞^6 divided by ∞^3, which equals ∞^3. A linear congruence is indeed a wonderful thing! Not only does it consist of ∞^2 lines closely knit in community with one another but it is woven through by infinitudes of infinitudes of quadric surfaces. It contains ∞^3 such surfaces. Later we will come to study some of these infinitudes of quadrics and to see the extraordinary beauty which they contain.

24.7 The ellipsoid and hyperboloid

It is now time to examine the various forms which quadric surfaces may have. The *sphere* and *spheroid* are obvious cases of surfaces which meet any plane in conic sections, and they therefore come under the heading of quadrics. The latter has elliptic sections on all planes except those which are at right angles to its major axis; these have circular sections. If we imagine the spheroid to be 'sat on' so that even these planes have elliptic sections, then it would be the *ellipsoid*, having elliptic cross sections only.

Next let us consider an ordinary hyperbola, with its central lines (Figure 24.4). If we imagine this curve to be rotated about the hori-

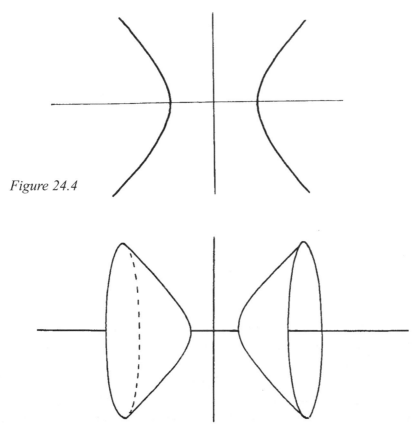

Figure 24.4

Figure 24.5

zontal axis we obtain what is called the *two-sheeted hyperboloid of rotation* (see Figure 24.5). In fact, we know that these two sheets are not really two, but join up through infinity, in a similar way to which the 'ends' of the plane hyperbola join up but in our drawing we have cut the surface off by two vertical planes in order to better portray it. The name, however, is a convenient way to distinguish it from the one-sheeted hyperboloid, which is formed by rotating this curve around the vertical axis. The *one-sheeted hyperboloid* is the same surface which we illustrated in Figure 24.3 (page 255), and is the form which is most readily obtainable, with its generators, by purely projective methods. It is the form with which we shall be most frequently concerned.

Notice that with the ellipsoidal types (sphere, spheroid, ellipsoid), as soon as we have adopted a viewpoint in space (either inside or outside the form) we can say quite definitely that, from our point of view, the form is concave or convex. Such a surface is said to be one of *positive curvature*. With the one-sheeted hyperboloid the case is different. Suppose that we adopt a viewpoint outside the form, somewhere more or less level with the waist; if we view the form along the horizontal plane which passes through our eye-point, it appears to be definitely convex; but if we view it along a vertical plane through our eyepoint, it is quite clearly concave. In between these two directions there are two in which there is no curvature at all; these are the directions in which the surface is straight; that is, the directions along which the generators run. Such a surface is said to possess *negative curvature*.

24.8 The paraboloid

Now let us take a plane parabola, with its central axis, as shown in Figure 24.6. If we rotate this figure about its central axis we get a *paraboloid of rotation*, or, as it is often called, an *elliptic paraboloid*. It is a surface of positive curvature and the plane at infinity is a tangent plane to the surface. If it is a paraboloid of rotation every plane at right angles to the central axis will meet it in a circle, but the general elliptic paraboloid will meet all such planes in ellipses.

Next let us consider a one-sheeted hyperboloid, as shown in Section 24.5. We can imagine the left side of the form to remain stationary, while we let the right side move out towards infinity. We can see how it would be possible for one of the points of the surface,

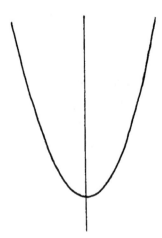

Figure 24.6

with its two generators, to go out to infinity, and for the plane of the
generators, which is the tangent plane at that point, to merge into the
plane at infinity. In such a case the form becomes a *hyperbolic
paraboloid.* It is a surface of negative curvatue and is crossed by two
generators at every point, just as is the one-sheeted hyperboloid. All
planes in a certain direction meet it in parabolas and all others in
hyperbolas.

Take again our two skew lines a and a', and put projective growth
measures on them between double points X and Y on a, and X' and Y'
on a'. If we now take A and A' to be the midpoints of XY and $X'Y'$, we
shall, of course, determine a quadric. By joining up the points of the
growth measures in order from A and A' we shall be able to form the
surface. Remembering that the growth measure in the external section
of the line is everywhere harmonic to that in the internal section we
shall see that one generator of this surface will be the line connecting
the points at infinity of a and a', that is, one generator will be a line at
infinity. Every point of this line at infinity will be paired with one of
its planes as tangent plane to the surface at that point, and one of these,
of course, will be the plane at infinity itself. Therefore, such a surface
would necessarily be a hyperbolic paraboloid. From this we can get the
proposition that if any three generators of one regulus of a quadric

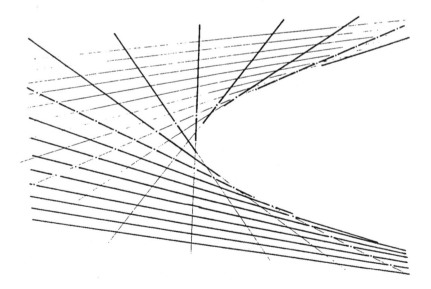

Figure 24.7

meet *two* generators of the other regulus in equal spacing, they must meet *every* generator of that regulus in equal spacing, and the quadric must be a hyperbolic paraboloid.

This leads to a celebrated construction for the hyperbolic paraboloid. Take any skew lines and mark equal intervals along them. Join any one point of the first line to any one of the second. Now continue joining points in order from the first pair. The lines so formed will lie on a hyperbolic paraboloid. (We have step measures along the two lines, and all step measures are projective with one another.)

The well-known construction for a plane parabola (joining equally spaced points along two lines) is really a plane representation of this three dimensional saddle-like form. It is drawn, in Figure 24.7, so that those lines which come in front are in thick, and those which pass behind are thin. This, of course, is only the one regulus of the surface; to represent the other we should merely have to change the thick and the thin lines. For the sake of clarity the evenly-spaced points from which the construction started have been shown darkly.

Since, in our studies, we shall find the surfaces of negative curvature being used more frequently than those of positive, in the future the word *hyperboloid* should always be read to mean *one-sheeted hyperboloid* and the word *paraboloid* to mean *hyperbolic paraboloid*. Where the two-sheeted and elliptic varities are intended it will be specifically stated. Unless otherwise stated, we shall be considering hyperboloids and paraboloids of circular cross-section. Analogous rules can be worked out for forms of elliptical cross-section.

24.9 Pole and polar

The laws of pole and polar for the quadric are analogous to those for the conic, and we state them here without further proof. They are illustrated in each case for the sphere, but the same laws hold for all quadrics whatever.

First we imagine a quadric with any point P inside it (Figure 24.8). Through P we put any plane, π', and of course this will meet the quadric in a conic. Now we imagine all the tangent planes of the quadric along this conic; they will form a cone. Question: if the plane through P starts to turn, first in one direction and then in another, always passing through P, what happens to the apex-point,

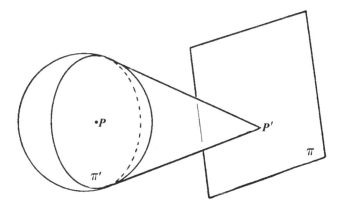

Figure 24.8

P', of this cone? Answer: it will move, first in one direction and then in another, across the surface of a plane. This plane, π, is the *polar plane* of P and P is the pole of π. Any line through P will meet the quadric in two points which are harmonic with respect to P and the point where the line meets π. The apex-point of the cone, P', is the pole of the plane π', through P, and π' is the polar of P'.

If we next imagine P to move gradually closer to the surface of the quadric we can see that its cone will grow more and more obtuse-angled, until when P is very near the surface, its cone will be nearly flat. As P enters the actual surface itself, its cone flattens until it merges into the tangent plane at that point. A point and its tangent plane are pole and polar.

All relationships of incidence are fully retained. Three planes passing through one point will transform into three points lying in one plane, etc.

Lines transform, of course, into lines. The common line of two points will transform into the common line of their polar planes. Two such lines are called *conjugate* or *polar lines*. The easiest way to find the polar line of a given line is shown in Figure 24.9.

Line h meets the quadric (in this case a sphere, represented by a circle) in points A and B. Their tangents are α and β, and therefore the polar, planes of A and B. Their common line, h', is the polar line of h. Since h' lies in the polar planes of all the points of h it is clear that every line of the congruence hh' meets the quadric in a pair of points which are harmonic with respect to the points in which it meets h and h'. It follows that if we take any plane of h', it will meet

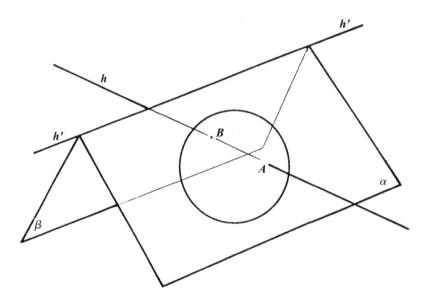

Figure 24.9

h in a point and the quadric in a conic, and this point and *h'* will be pole and polar with respect to the conic.

24.10 Projectivity on polar lines

Now let us take a hyperboloid and cut it in two real points, *S* and *T*, by a line *z*. We imagine the tangent planes at *S* and *T*, and their common line *m*. We let *m* meet the hyperboloid at *M* and *N*. Lines *z* and *m* are now polar lines with respect to the hyperboloid (see Figure 24.10). But then the two generators through *S* must pass through *M* and *N*, since these generators determine the tangent plane at *S*; that is, the polar plane of *S*, and since *S* lies in *z*, this polar plane must lie in *m*; and the same is true for the generators through *T*. *SM* and *TN* are generators of the one regulus and *TM* and *SN* are generators of the other. Let us consider now any other generator, *g*, of the other set. Line *g* will then meet *SM* and *TN* but will be skew to *TM* and *SN*. If we imagine all the lines of the regulus determined by *m*, *g* and *z* we know that they will make projective ranges along *m* and *z*, and we can see that two of these lines are already drawn for us: *SM* and *TN*. It follows therefore that in this projectivity *S* corresponds to *M*, and *T* to *N*. If we put a growth measure of a certain cross ratio along z, with double points S and T, we shall get a corresponding growth measure, with the same cross ratio along m and its double points will be M and N.

Notice that with a hyperboloid, if a line meets it in real points its polar line must also do so. Had z met the surface in imaginary points so would m, and we should have had to work with circling measures instead of growth, but the essential principle would have been unchanged.

24.11 An important case of this projectivity

Now let line z become the horizontal line at infinity. Where shall we find its polar line, m? Take two planes of z — the horizontal planes at the top and bottom of the model. They meet the hyperboloid in circles* and the poles of z with respect to these circles are their centres. Therefore line m must be the central vertical axis of the hyperboloid.

In what points does the hyperboloid meet line z? Notice, in Figure 24.10, that any plane of z would meet the surface in a conic and S and

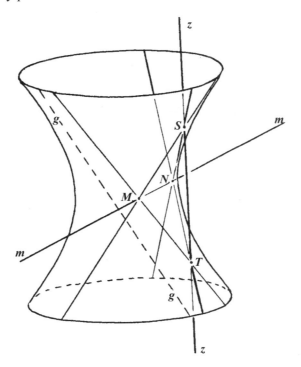

Figure 24.10

T are the common points of *z* with this conic. Therefore, it is clear that our new *z*, the horizontal line at infinity, meets the hyperboloid in the points *I* and *J*.

Let us take any generator, *g*, of the hyperboloid, and consider all the lines which meet *m*, *g* and *z* (Figure 24.11). They will consist in fact of all the horizontal lines of *m* which meet *g*. Since they are all the lines which meet the three skew lines *m*, *g* and *z*, they will form a regulus and since one of the skew lines is at infinity this regulus will lie on a paraboloid. Let us imagine a horizontal line of *m*, rising up the axis, and always turning so that it passes through *g*. It will have a spiralling motion which will sweep out this paraboloid.

Now let us put a circling measure along *z*, between double points *I* and *J*. We know that this will consist of equiangular steps — let us say 10°. By the above we know that the horizontal line through *m* will have to move along *m* projectively with this — that is, it will move in a circling measure along *m*, and will go 10° along this circling measure for every 10° turned.

This is a most important property of the hyperboloid, and is much used later on in our constructions. By it we can now tell at what

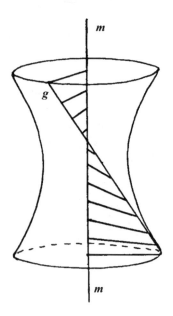

Figure 24.11

* See the end of Section 24.8.

points the central axis meets the hyperboloid: they will be the double points of the circling measure which this horizontal line makes along the axis.

The amplitude of a circling measure we know is marked by the points which are 90° apart, and equally spaced on either side of the centre. If we imagine our hyperboloid to be cut off by horizontal planes at such positions equally above and below the central waist that each generator meets these planes at points which are 90° apart, then these planes mark the amplitude of the imaginary points in which the hyperboloid meets the central axis.

If we have a central axis containing the imaginary points M and N, and we wish to construct a hyperboloid passing through M, N, I and J, we have to put two planes through the amplitude of M and N, at right angles to line m, and in these planes we put equal circles. Round these circles we make an equi-angular measure. Now if we join points in the top circle to points in the lower one which are 90° away, we shall be making the hyperboloid which we want.

24.12 Degeneration

We will close this chapter by considering two important degenerations of the quadric. In Section 6.6, we saw how a conic can degenerate into a pair of lines, and an analogous thing can happen to the hyperboloid. Suppose we have a hyperboloid containing the imaginary pair MN and that the amplitude of these is the distance AB (Figure 24.12). We will make the radius of the circles equal to AC.

We would now have to make our generators go from C to D, and so on round the circles. But if we were to keep these circles as they are and let the amplitude of the pair MN get smaller, we should find that the distance between the circles would now represent more than 90° in the circling measure of MN. Therefore, we would have to draw our first generators more like line CE, and the hyperboloid would become narrower waisted. We know that decreasing the amplitude is equivalent to letting the imaginary points M and N draw closer together. Our hyperboloid would now look something like Figure 24.13.

If we let this process continue to the limit we shall be joining points on the upper circle to those on the lower one, 180° apart.

All these lines will meet at the centre and our hyperboloid will have degenerated into a cone. At the same time the amplitude of the

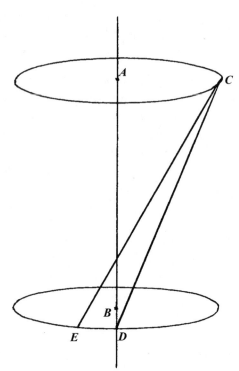

Figure 24.12

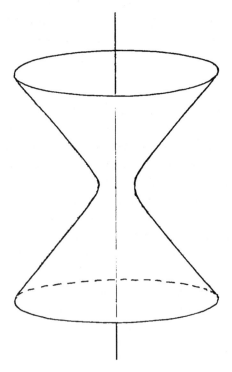

Figure 24.13

pair *MN* will have become nil, they will have met, and have become
the real pair coincident at the apex of the cone. Simultaneously, the
pair of generators which go through each point of the hyperboloid
will have each become coincident, and will form the generators of
the cone. Looked at from this point of view the cone, as a degenerate quadric is formed of a coincident pair of points through which
go an infinitude of coincident pairs of lines.

On the other hand we could let the amplitude grow larger than the
distance between the circles. This distance would then count for less
than 90° and we should have to join the generators up much more vertically. In the limit the hyperboloid would become a cylinder; this
would be when the amplitude would be infinite; that is to say when *A*
and *B* have become coincident at infinity. There they would become a
real coincident pair, and would in fact be the apex of the infinite cone
which a cylinder is. These two degenerations are really just two
aspects of the same thing.

But we could keep the radii of the two circles, and the 90° joining
up, the same, but bring the two planes *AC* and *BD* closer and closer
together. When they were very close they would look like Figure
24.14.

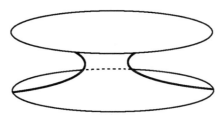

Figure 24.14

In the limit, when the two planes coincide the hyperboloid will have
become a line-wise conic in that plane, each line being again formed
of two coincident generators.

25. The Imaginary Part of a Sphere

25.1 Pole and polar relationships

Let us consider a sphere and any plane, α, which does not meet it in real points (Figure 25.1). Corresponding to α within the sphere will be its pole A. Let B be any point of α. From B there will be a tangent cone to the sphere, and this will touch it in a plane β, which will be the polar plane of B. But plane β will meet plane α in a line, b. The presence of the sphere has therefore caused a certain correspondence between point B and line b.

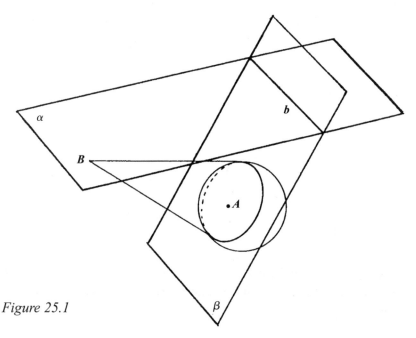

Figure 25.1

Let us draw the perpendicular from B to b, and now we can take any other point of this perpendicular, and find its corresponding line in a similar way. We shall find that it is parallel to b. In Figure 25.2 we shall show this configuration in section by a plane through the perpendicular and the centre of the sphere.

The sphere will show as a circle, and α and A will show as a line

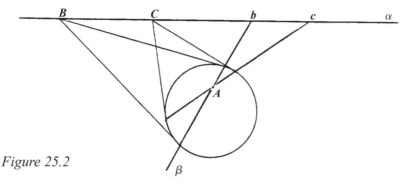

Figure 25.2

and point which are polar and pole with respect to it. The cone from *B* will show as two tangent lines to the circle, and the plane *β* will show as the polar line of *B*. Line *b* in plane *α* will show as point *b* in line *α*. If now we take a new point *C* and repeat the construction we will have a figure which is familiar to us. It is to be found in Figure 15.3 (page 165). *B* and *b*, *C* and *c* etc. are pairs of points in circling involution along line *α*.

25.2 The imaginary circle

But now if we return to our full, three-dimensional view, Figure 25.1, we have points and lines paired in correspondence along a circling involution, like Figure 25.3.

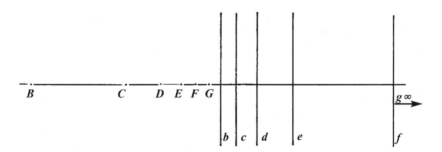

Figure 25.3

This also is something which is familiar to us. It is the complete and exact pole and polar correspondence of an imaginary circle (see Section 18.3). *B* and *b*, *C* and *c*, are pole and polar with respect to an imaginary circle, and the real conjugate radius of this circle (that is, the radius which the circle would have had if it had been real) is the distance of the amplitude of the circling involution on which the points and lines lie.

We have said that a quadric surface is one which meets all planes in conics and now we see the universality of this statement. A plane which meets the sphere in real points will do so in a real circle. But if it meets the sphere in imaginary points it will do so in an imaginary circle. One cannot of course see that circle but it reveals its presence by the fact that it has set up its characteristic pole and polar correspondence within the points and lines of the plane, just as an invisible magnet held under a piece of paper reveals its presence by the patterns which it induces in iron filings which have been scattered on top. This truth works analogously for all other quadric surfaces.

25.3 The imaginary part of a sphere

Now we will follow a similar line of thought to that followed in Section 15.3. There we have a Figure (15.5) showing what happens to the amplitude points of the imaginary pairs, as the line moves parallel to itself towards and away from the circle. The imaginary pairs themselves remain invisible but we take the real points which mark their amplitude and watch what happens to them. We saw that they follow a rectangular hyperbola, which is tangent to the generating circle.

In the case of the sphere, instead of a pair of amplitude points we have the real conjugate circle. This is the real circle which shows the 'size' of the imaginary one in question. The radius of the imaginary circle is i times that of the conjugate circle.

The sphere meets the plane in an imaginary circle, and this circle is represented for us by the conjugate circle which we can draw. What happens to this conjugate circle as the plane in which it has been engendered moves parallel to itself, towards and away from the engendering sphere? The answer is the one which we would expect by analogy from the circle. The conjugate circle traces out a two-sheeted hyperboloid as illustrated in Figure 24.5.

25.4 The absolute circle

The sphere is surrounded by ∞^2 such hyperboloids, one for every *direction* of planes in space. If one follows these planes outwards in space, each one bearing its constantly expanding imaginary circle, one has to ask what happens when all these planes merge into one, in the plane at infinity. The answer is that all the imaginary circles merge into one vast imaginary circle, which lies in the plane at infinity.

In considering this circle we come to a matter of utmost importance. It is called the *absolute imaginary circle at infinity* or sometimes, for short, the *absolute circle*. It is the counterpart for three-dimensional space of the points *I* and *J* for the plane. In Section 15.5 we dealt with *I* and *J*, and showed how these are the true controlling elements for the metrical properties of Euclidean space. We were able to define parallelism, right angles, equal intervals, circles, etc. all in terms just of the infinite line, and the points *I* and *J* within it.

In three-dimensional space we can do the same thing with reference to the plane at infinity and the absolute circle within it, for instance:

1) Two points which are such, with respect to a conic, that each lies on the polar of the other, are called *conjugate points*.
2) Two lines of space which meet the absolute plane in points which are conjugate with respect to the absolute circle are at right angles.
3) Two planes which meet the absolute plane in conjugate lines (which lie on one another's poles) with respect to the absolute circle are at right angles.
4) Any quadric surface which meets the absolute plane in the absolute circle is a sphere.

If we ask what the absolute circle is like we can say, 'It is that circle in the infinite plane which makes every equator, polar to its pole.' If we look at the heavens, realizing that they are really a flat plane, we see that the celestial equator is really the representation of a straight line. It is the polar line of the celestial pole with respect to the absolute circle. If we now imagine the equator to tilt about one of its points (this point will be seen in the heavens as two points 180° apart) then its pole with respect to the absolute circle will move so that it is always just 90° away from the equator.

The sum total of all such pole and polar correspondences in the infinite plane is the pole and polar relationships of the absolute circle. A circle can also be looked on as the sum total of all its points. The absolute circle is the sum total of all the *I*s and *J*s of all the infinite lines there are.*

---

* Every plane, including the one at infinity, has ∞^2 lines. So there are ∞^2 circling points, which equals the number of points on any imaginary circle.

25.5 Imaginary generators

This is the conception which we come to when we follow the imaginary part of a sphere out towards the infinite. Now we must follow it inwards towards the surface of the sphere. As we do so the radius of the conjugate circle becomes progressively less until when the plane is tangent to the sphere it becomes nil. Let plane α be tangent to the sphere and, of course its polar, A, will now be in it and will be its point of contact with the sphere, Figure 25.4. Now we take any point B of plane α, and draw the tangent cone to the sphere. It is obvious to the imagination that plane α will be tangent to this cone, that the cone will touch the sphere in a plane which will meet plane α in line b which will pass through point A, and will be at right angles to line BA.

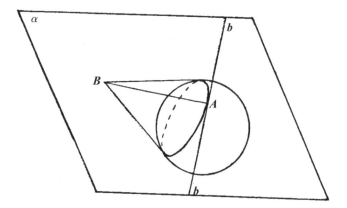

Figure 25.4

It is clear that pole and polar relationships in the ordinary sense have ceased to exist, since every point of the plane now has its polar passing through A. The relationship has become a right-angled involution in point A. Every point of the plane has its polar passing through A at right angles to the line which joins it to A. The quadric engenders in plane α and point A a right-angled involution which represents a pair of imaginary lines; the conic in which the sphere meets plane α has degenerated! This pair of imaginary lines through A is in fact the pair of generators of the surface through that point. We see that all quadrics whatever have two generators through each point. If the generators are real the surface is a saddle-like, waisted one, of negative curvature; if they are imaginary it is one of positive curvature.

Let us take two tangent planes to the sphere, α and β, with their points of contact A and B (Figure 25.5).

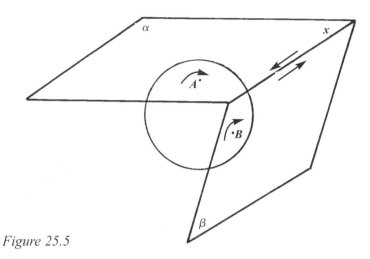

Figure 25.5

We will assume the convention that all points I have the clockwise rotation, and points J have the anti-clockwise. Planes α and β meet on line x and we can see from elementary geometry that x will be equidistant from A and B. The imaginary generators in A and B will therefore meet line x in an identical involution. We have marked in our figure, the clockwise rotation of the lines in planes α and β, relative to the viewpoint from which the sphere is seen in our figure and the circling motions which they engender in line x. We see that the two lines with the clockwise rotation do not meet; and just as obviously the anti-clockwise line of A meets the clockwise line of B, and vice versa. We therefore see that these generators lie in two sets, one clockwise and one anti-clockwise and that every clockwise line is skew to every other clockwise one. We see thus that these generators follow exactly the same kind of relationship as the real ones which we can see on the surface of a hyperboloid.

26. The Twisted Cubic

26.1 The meet of two quadrics

We have seen in Section 23.4 how we can consider a twisted curve; it is a single infinitude each of points, lines and osculating planes in space; whereas a surface is formed of ∞^2 points and tangent planes. The order of a twisted curve is given by the number of points in which it meets any arbitrary plane of space; whereas the order of a surface is given by the number of points in which it is met by any arbitrary line, or, what comes to the same thing, it is the order of the plane curve in which it is met by any plane.

Now let us consider any two quadric surfaces. They will have points (and planes) in common, and these will lie on some locus. Let us consider their common points to begin with. We will cut the two quadrics by some arbitrary plane of space. Each quadric will meet the plane in a conic section, and we know that two conics always have four common points, real or imaginary (Chapter 12). Clearly these four common points are common points of the two quadrics, and we thus see that the common points of these two quadrics lie on a locus which meets any plane of space in just four points. In other words the two quadrics meet on a twisted curve of fourth order — a twisted quartic.

An interesting family of such curves can be imagined, and drawn, fairly easily, using the unprojective method of *plan* and *elevation* (see also Section 29.1). We will take for our two quadrics, one a sphere and the other a degenerate quadric, a cylinder, and we will let these two meet. The three drawings of Figure 26.1 represent

(1) a *front elevation* or *picture plane* — the circle represents the sphere and the rectangle represents the cylinder, cutting through the sphere in the front part of it;
(2) a *side elevation* and
(3) the same thing seen in *plan*, the large circle representing the sphere and the small one the cylinder as seen above.

We now imagine a horizontal plane, anywhere we like, but cutting the sphere. This plane is represented in the left and middle figures

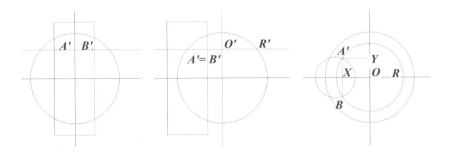

Figure 26.1

as a horizontal line. The distance $O'R'$ in the middle part of Figure 26.1 shows the radius of the circle in which the plane meets the sphere, and this can be transferred to the circle through R in the right part of the figure. This circle meets the small cylinder-circle at A and B. The distance OX can now be transferred to the middle figure giving $A'=B'$. This is the point in the side plane in which cyclinder, sphere and horizontal plane meet. Similarly OY must be transferred to the left figure giving A' and B', and these are the meeting points on the front elevation. One can now repeat this process for a number of horizontal planes, and so plot the curves in which the surfaces meet, on the two elevations.

It is most interesting to do this with the relationship of sphere and cylinder as they stand in Figure 26.1; then to move the cylinder inwards a little, until it shares a tangent plane with the sphere; that is, until the two circles in Figure 26.1 have a common tangent; and then to let the cylinder move even further inwards, so that it passes completely inside the sphere; that is, the small circle is wholly within the large one. In Figure 26.2 we have done this for the first case; in Figure 26.3 for the case where the surfaces have a common tangent plane, and in Figure 26.4 for the case in which the cylinder passes wholly through the sphere.

We see that we get a spatial set of curves which are the equivalent of the 4th order family which are known as the *curves of Cassini* in the plane. They go all the way from the oval, through the dumb-bell to the

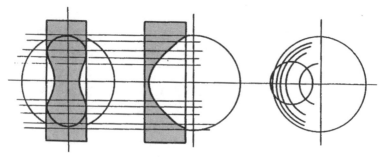

Figure 26.2

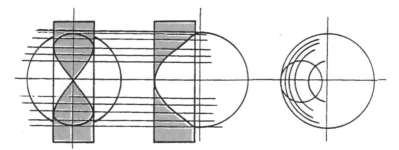

Figure 26.3

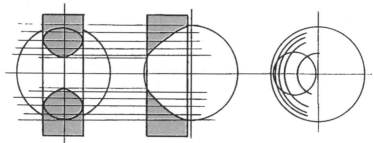

Figure 26.4

lemniscate, after which they separate into two 'eggs.'* We studied the
plane curves of this family in Section 20.4. Similar things could be
done with other quadrics, say two hyperboloids, etc. but the construction methods would be much harder.

* A true Cassini curve is, of course, two dimensional. There is, however, a significant difference with the Cassini curves. After the 'lemniscate' the curves
  become smaller and degenerate into a point when cylinder and sphere touch
  outwardly. Also when the axis of the cylinder contains the centre of the sphere,
  the two 'eggs' become circles that appear as stripes in our picture.

Exercise 26a

Work out, and use, a similar construction for two cylinders, intersecting at right angles.

26.2 The twisted cubic

Now we must consider two quadrics which have one generator in common. They will meet along this generator and also along some curve. Any plane of space will still meet them in two conics, with their four common points, but one of these points will be taken by the common generator. The rest of the meeting points must therefore lie on a curve of third order — a *twisted cubic*. In fact it is found that this is the lowest order space curve which one can have; as soon as a curve becomes second order it automatically becomes a plane curve — in fact a conic. In the world of space curves therefore the twisted cubic is the simplest which we can find, and it is for this reason one of the most fundamental and important curves to study. It has a position in the world of space curves which bears many resemblances to that which the conic occupies in the world of plane curves.

26.3 The meet of two cones

Firstly we will study how to draw perspective pictures of twisted cubics, considering them as the meeting curve of two quadrics which have a common generator. These quadrics may be degenerate, for instance, cones, the apex of each of which lies on the surface of the other.

Let us consider a second order cone. It need not be a circular one; its right section may be an ellipse or a hyperbola. We may consider it linewise; that is, as being made of all its lines passing through the apex; each of these lines is a generator of the cone. If the apex of each lies on the surface; that is, on a line of the other, the line joining their two apexes will be their common generator, x. Any plane of x will meet each cone in just one other of its generators; these two generators being coplanar will have a common point. and this will be a point of the cubic.

Now let the model be cut by any plane of space. This plane will meet it in two conics, α_1 and α_2, with their common points, and through one of these, say D, the common generator must pass (see Figure 26.5).

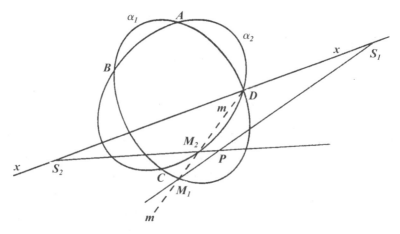

Figure 26.5

At this point we must transform the figure into a three-dimensional one in our imagination. Because the conditions of our problem have not yet been enough to fix our viewpoint we can do this in a variety of ways, each of which is as correct as each other. We must establish some conventions which will sufficiently define our viewpoint. For instance we will say that both conics lie on the horizontal plane of the page, and that the 'lower' part of the figure is in front, and the 'upper' part, behind. That is to say that C is close to us, B is farther away and A is farthest of all. Now we draw through D a line, x, to represent the common generator. This line does not lie in the plane of the page; we will suppose that it comes upwards as it apparently passes from left to right. We mark on it the apexes of the cones, S_1 and S_2. This will mean that S_2 is below the page and S_1 is above. We will let the S_1-cone meet the horizontal plane in conic α_1, and the S_2-cone in conic α_2. Now if we join S_1 to any point of α_1 we will have a line of the S_1-cone. Similarly S_2 joined to any point of α_2 will give us the S_2-cone. Neither of these lines will lie in the plane of the page; although they will *appear* to meet on the page, they will, in general, be skew, passing over and under one another in space, and never meeting. Our problem is to find generators of S_1 and S_2 which meet. It is easily solved.

Let us consider any plane of the common generator. It will meet the page-plane in a line, and clearly this line must pass through D. Therefore we draw any line through D, say m, and we let this meet α_1 and α_2 in M_1 and M_2. S_1M_1 and S_2M_2 are generators of the cones, and they share a common plane; therefore they have a common point; P is a point of the cubic. By turning the line m around D we can plot as many points of the cubic as we wish.

Notice that because P falls outside the section of the lines S_2M_2 it must lie above the page, and similarly because it lies inside the section S_1M_1. As m turns about D in a clockwise direction P will be found to move towards C. When it reaches it, it will be on the page, and thereafter, for a time, P will fall between S_2 and M_2, and it will be outside S_1M_1, by both of which tokens we will see that it will be below the page. Later it will come up through the page again at B, and later still will go down again through A, thus meeting the horizontal plane in three points as a cubic should.

It is exceedingly instructive to make a series of figures of this nature. The ellipses can meet in four points or in two; one, or both, of them may become a hyperbola, giving a hyperbolic cone; the angle of the common generator through D may be varied ... There are infinite possibilities. While doing this the three-dimensional reality of what one is drawing should be kept vividly in mind the whole time; then it will be clear which part of which cone obscures, or is obscured by, the other. Shading the obscured parts lightly, and the unobscured darkly, one can begin to build up a clear idea of what a twisted cubic looks like.

In the present figure, for instance, the part of the S_2-cone which lies from P towards M_2 is obscured and the part which lies from P away from M_2 is showing. But the part of the S_1-cone which lies from P towards S_1 is obscured and the part which lies towards M_1 is showing.

26.4 The meet of two hyperboloids

We are now in a position to extend the construction of Section 26.3 to include the meet of two hyperboloids having a common generator. First we must remember an important quality of the hyperboloid. If we take any generator of the one set, and consider any plane of that generator, that plane will contain just one generator of the other set (Section 24.4).

We will now take two congruent and similarly placed ellipses (Figure 26.6). We can consider these as being placed on two horizontal planes of a model, the upper one on the top plane and the lower one on the bottom. (As a perspective picture this will of course mean that the lower ellipse is actually larger than the upper one, but

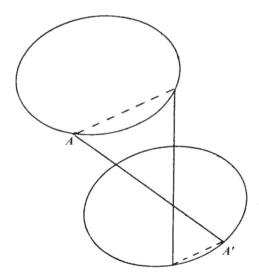

Figure 26.6

being of the same eccentricity it will meet the line at infinity in the
same pair of imaginary points. Our figure is therefore still valid for
making a hyperboloid).

We will now take any point, A, of the upper and join it to any point,
A', which we wish, of the lower one. This will give us a generator of the
one set. Next we take any plane of this generator; it will meet the top
and bottom plates of the model in a pair of parallel lines (dotted). This
plane will contain one generator of the other set, which must obviously
join the other points in which the parallel lines meet the conics.

This is another, and very convenient, way of constructing a hyperboloid. By turning the plane around the generator one can construct as
many positions of the other set of generators as one wishes.

Now we can do the whole thing with two conics on each plate, as in
Figure 26.7. The first generator must go from a common point on the
top plate to a common point on the bottom. This then is the common
generator.

Now we take any plane of this generator, represented by a pair of
parallel lines, and we can construct generators of the other set for each
conic. These generators, sharing a common plane, must also have a
common point. They are not skew, but meet at point P. This point is a
point of the cubic. By turning the plane around the common generator
one can construct the cubic with its interpenetrating hyperboloids. As
an exercise this is very well worthwhile doing, and in the doing of it
one gains a very good idea of the way in which the cubic weaves

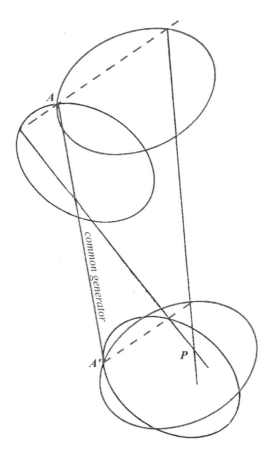

Figure 26.7

between its two hyperboloids. It is not, however, easy, and it is very hard indeed to portray both the hyperboloids with their intersecting curve on one plane figure. We show in Figure 26.8 the result of work-ing Figure 26.7, showing the curve only on one hyperboloid, which comes out to be, in this case, almost cylindrical. The whole thing is obviously well adapted for a string model and it should be studied in this way.

Figure 26.8 is well worth studying carefully. All light lines and dot-ted curves are at the back. Notice how the curve comes up from below, on the front of the hyperboloid, meeting the lower horizontal plane where the two conics cross, at *A*, thereafter soaring upwards in a grace-ful curve towards the left; then downwards again, passing to the back of the hyperboloid, it crosses the lower horizontal plane, at *B*; and after a long sweep downwards across the back of the hyperboloid it finally turns upwards again and coming round to the front it crosses the lower horizontal plane for the third time at C, thereafter making its long journey upwards, through the top horizontal plane at D, and so to infinity, where it will join up with its 'other end.' Also there is the common generator and we should notice that, counting this as well as the curve, every common point of the four conics is accounted for. One has to understand, of course, that the cubic also contains two imaginary points on the top horizontal plane.

26.5 Two projective stars of lines

Now we must approach the matter from a different point of view, and one which will enable us to attain a remarkable amount of information about these curves by applying just the elementary first principles of projective geometry. Let us imagine any two points of space, S_1 and S_2. They will each contain ∞^2 lines — a *star of lines*. In Section 6.1, we learnt how we could establish a projective, one-to-one relation between the lines of two plane pencils, so that to each line of the one pencil there would correspond one unique line of the other. We found that common points of corresponding lines all lie on a conic.

Now we can equally envisage a one-to-one relationship between the lines of the two stars in S_1 and S_2. To every line of S_1 will correspond, in some projective, one-to-one relationship which we have the freedom to fix in any one of a multitude of possible ways, one unique line of the star in S_2. We can now ask the equivalent question to that which we studied in Section 6.1: where do the meets of corresponding lines lie? Things are a little more complicated here, because if we take any line of S_1 and its corresponding line of S_2, we have no guarantee that they will meet at all — they may be skew. In fact, generally they will be! But we can surely have confidence that at least a few of them will meet, and these are the ones we are interested in to begin with.

Now let us consider any arbitrary plane of space. Every line of S_1 will meet it in just one point; there is a one-to-one (in this case, a direct perspective) relationship between the ∞^2 lines of S_1 and the ∞^2 points of the plane. Similarly with the lines of S_2. Therefore, a one-to-one relationship between the lines of S_1 and S_2 will engender a one-to-one relationship between the points of the plane. Any two corresponding lines which are skew will meet the plane in a pair of corresponding

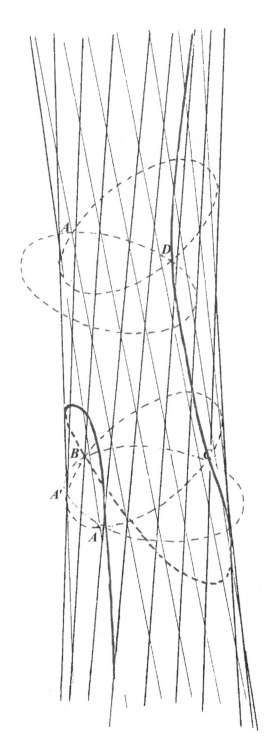

Figure 26.8

points which are distinct, but a pair of corresponding lines which meet *on the plane* must meet in a self-corresponding point of the plane projectivity. In Section 21.1, when studying path curves, we found that any plane projectivity always has three self-corresponding points — the invariant triangle.

Thus we now know what is the locus of the meets of corresponding lines in the two stars, S_1 and S_2; it is a curve which meets every plane of space in just three points. Out of the ∞^2 pairs of corresponding lines in S_1 and S_2 only ∞^1 will meet, and they will lie on a twisted cubic. This curve obviously contains S_1 and S_2.

We now begin to see something of the truth of the statement that in the world of space the twisted cubic occupies a similar position to that of the conic in the world of the plane. And this truth leads us to a path of enquiry which will show us that the twisted cubic, taken in its entirety, is an 'organism' of most amazing and wonderful complexity.

We refer back to Sections 24.2 and 24.3. There we generated a quadric by moving the point M along line b, between the two skew lines, a and c. For every position of M there was a plane of a and a plane of c, and the common line of these two planes was a generator of the quadric. This was equivalent to saying that a quadric is generated by the common lines of corresponding pairs of planes in two projective pencils of planes, provided the pencils are based on skew lines.

Now as soon as we determine a one-to-one relationship between the lines of the stars of S_1 and S_2 we automatically determine a similar relationship between the planes of their sheaves. Consider any two lines of S_1. They will have a common plane, and they will correspond to two lines of S_2, which have *their* common plane. These two planes will correspond in the projectivity between the sheaf of planes in S_1 and the sheaf in S_2. And unlike the lines, of course, every pair of corresponding planes will meet — in a line.

Let us consider such a pair of corresponding planes, α_1 and α_2. They meet in line x (see Figure 26.9). Each line of S_1 which lies in α_1 must correspond to a line of S_2 lying in α_2. Thus, the pencil of lines S_1 in α_1 is projective with the pencil of S_2 in α_2. The lines of S_1 will meet x in a range of points which will be projective with the range of points in which the lines of S_2 meet it. Now we know that two projective collinear ranges always have just two self-corresponding points. Thus, line x must meet the cubic in two points.

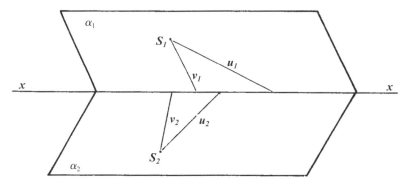

Figure 26.9

It is possible for a line of space to meet a twisted cubic in no points, in one point or in two. It cannot meet it in three unless the cubic is degenerate. A line meeting the cubic in one point is called a *unisecant* of the cubic, and one meeting it in two points is called a *bisecant*.

We now see that x, the meeting line of any two corresponding planes of S_1 and S_2 is necessarily a bisecant of the curve. Next let us take any pair of corresponding skew lines of the pencils $S_1\alpha_1$ and $S_2\alpha_2$. We mark them as v_1 and v_2 in Figure 26.9. The pencil of planes in v_1 will correspond projectively with the pencil in v_2, in the general projectivity S_1 to S_2, and we know that common lines of corresponding planes will lie on a quadric; we have also found that each of these lines must be a bisecant of the curve. Thus the cubic lies on a quadric, all the generators of one set being bisecant to the curve, and amongst these generators, of course, is line x of our figure.

Now consider any one of these generators, say line x. Through each point of x will pass one generator of the other set. Let us consider any one of these, say x'. Lines x and x' determine a plane, which is in fact the tangent plane to the quadric at their common point. Since the cubic meets x in two points it must meet the plane also at these points. Since it is a cubic it must, and can only, meet this plane in one other point. Since the cubic lies on the quadric it must meet this plane in just one point outside line x *on the cubic*. But the tangent plane only meets the cubic along the lines of its two generators. Thus the third meeting point must be on the other generator x'. Thus we see that the twisted cubic must lie on its quadric in such a way that it meets every generator of one set twice, and every generator of the other, once.

Notice that x can only be generated from the planes of S_1 and S_2 by the meet of planes α_1 and α_2. No other pair will contain it. Considering now the quadric generated by v_1 and v_2, we take any corresponding pair of planes of these lines, and we see that they meet in some line y, skew to x. This line y can *only* be the meet of just these two planes of S_1 and S_2. Therefore if we now generate a quadric from some other pair of corresponding lines, say u_1 and u_2, of α_1 and α_2, this quadric cannot possibly contain line y. Thus it must be a completely new and distinct quadric from the one generated by v_1, and v_2, although it will share x with it as common generator.

Thus each pair of corresponding lines of S_1 and S_2 generates a distinct quadric on which the cubic lies. But there are ∞^2 such pairs!

We have thus far thought of the twisted cubic as being the meet of one pair of intersecting quadrics, but we now see that it is in fact the meet of ∞^2 such quadrics, marvellously interweaving, each quadric sharing just one common generator with each other, each one meeting the cubic twice with each of one of its sets of generators, and once with each of the other of its sets of generators, the whole forming an immensely intricate organism of which the cubic itself is the central golden thread. The imagination boggles!

26.6 Determining a cubic: six points in space

We know that a conic is determined by any five points in a plane, and we must now ask the equivalent question for the twisted cubic: how many points in space will completely determine a twisted cubic? The answer is six, and the proof for this is quite simple. Let us be given six points in space, no three of which are collinear and no four of which are coplanar. Let them be A, B, C, D, E and F. If we now choose F as a centre we can join it to the other five points. But we know that five lines in a point will determine a second order cone in that point. If we now join E to the other five points we shall have also determined a second order cone in E. But these two cones have a common generator, therefore they meet in a twisted cubic, and this is the unique curve determined by the six given points.*

---

* Alternative proof. Take E and F as centres of stars and connect them to the other four. This defines a projectivity between these stars, which in turn defines a cubic (Section 26.5).

26.7 Five points on a hyperboloid

We now have to ask how many points, on the surface of a quadric, are needed to determine a twisted cubic which shall lie completely on that quadric. The answer is five, but we shall find that, given any five points on a quadric, there are always just two cubics which will go through them all. We see that we can do geometry on the surface of a quadric, just as before we did it on the surface of a plane. On a plane, if any three of the five points which determine a conic are collinear, then the conic will be degenerate. On the quadric the same holds for the cubic if any three of the five points lie on one generator.

Now let us put five random points on a quadric, *A*, *B*, *C*, *D* and *E* (Figure 26.10). We will have *A*, *B* and *C* on the front (drawn thick) and *D* and *E* on the back (drawn thin).

We draw the line *AE*, piercing the front of the quadric and emerging from the back through *E*; we call it *u*. Now we take any generator of the one set, *g*. We can, of course, put the planes of *u* into one-to-one relationship with those of *g* by fixing the correspondence of any three

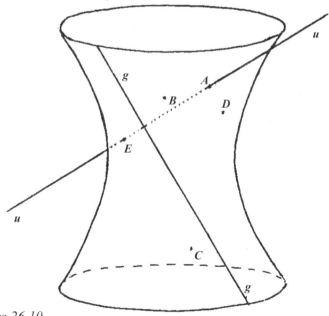

Figure 26.10

pairs of planes. So we can let the planes uB, uC and uD correspond with gB, gC and gD. This will form a regulus which we will call the ug quadric, having g for a common generator with our original quadric; therefore these two will meet on a cubic. Since uB corresponds with gB the ug quadric will have a generator going through B, and since the cubic is formed of the meet of this quadric with the original one, the cubic must pass through B. Similarly for C and D. Obviously u is also a generator of the ug quadric, so the curve also goes through A and E. Therefore, being given any five points on the quadric we have proved that it is possible to draw a twisted cubic through them. All the generators of the original set will be bisecant, and all of the other will be unisecant, to this cubic. We have yet to prove that this is the only possible one in this relationship.

For we could have done the whole process with some other generator of the original set, say g_1 or g_2. Would we then have come to the same, or a different, cubic? Now we must remember that each of the various planes of g (generator of the first set) contains just *one* generator of the other set. We must remember also the fundamental law which we studied (Section 24.3) that if we have any four generators of the original set, these will each determine a plane with any given generator of the other set, and these four planes will have a certain cross ratio, and that if we now go to any other generator of the original set we like, this cross ratio will be unchanged. It means that the planes of g are in exact one-to-one relationship with the lines of the other set of generators. (Since there are ∞^1 planes and ∞^1 lines, it is perfectly possible to have a one-to-one relationship between them).

Now since the planes of g are naturally in one-to-one relationship with the lines of the other set, and since we have put the planes of g into one-to-one relation with the planes of u, there must also be a one-to-one relation between the lines of the other set and the planes of u. We will call the relation of the planes of g with those of u, the ug relation, and the relation of the other lines with the planes of u, the ul relation.

Let us take any generator of the other set, say v. It will correspond to some other plane π of u, according to the ul relation, and it will correspond to its common plane with g in the gl relation. This common plane and plane π will determine a line which pierces the original quadric in a point of the cubic. But this common plane meets the quadric all along v, therefore the point of the cubic is the point in which v meets π.

Now if we take some other of the original set of generators than g, say g_1, and repeat the whole process, the ug relation will be changed to the relation ug_1, but all the cross-ratio relationships of the other set of generators will be unchanged, by the fundamental law quoted above. This means that the ul relation will be unchanged. Therefore v will still correspond to plane π, and will still meet the cubic in the same point. This goes for every generator of the other set, and therefore the cubic is unchanged.

If however we had chosen a generator of the other set, in place of the original one, for our original relationship with the planes of u, we would have got a different cubic, which would have all the other generators for bisecants and the original ones for unisecants. It would spiral in the other direction round the quadric.*

26.8 To construct a cubic through five points on a cylinder

These considerations lead to some very nice constructions: for instance, the twisted cubics which go through any five points on the surface of a cylinder.

Firstly we draw our cylinder in perspective, seen from above (Figure 26.11). It need not have a circular right-section, therefore we can, for convenience of drawing, put actual circles at top and bottom. The lower part of each circle represents the part that is nearer our eyes. On it we will place five points, A, B and C at the front, and D and E at the back. We join AE for our line u, and our first problem is to find the points in which u meets the top and bottom plates of the model.

This is easily done. For the vertical generators through A and E, being parallel, have a common plane and this clearly meets the top and bottom plates of the model in the lines e and e', as marked in the figure. But this common plane also contains the line u; therefore where e and e' meet u, are the points in which u meets the top and bottom plates. We call these points U and U'.

Now we take any further generator of the cylinder, g. We let there be a projectivity between the planes of g and the planes of u. This will produce the lines of a regulus which will meet the cylinder in a twisted

---

* For an introduction to cubics see Theodor Reye, *Geometrie der Lage*, Part II, lectures 21ff.

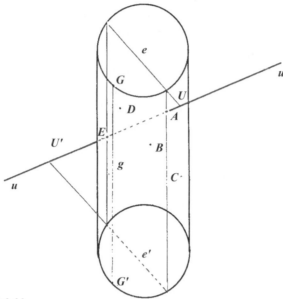

Figure 26.11

cubic passing through A and E. But the planes of g meet the top and bottom plates in pencils of lines through G and G'; likewise the planes of u meet the top and bottom plates in pencils of lines through U and U'. And all the lines of the pencil in G will be parallel with those of the lines of the pencil in G', etc. Since the planes of g are projective with the planes of u, it follows that the pencils in G and U will also be projective; likewise those in G' and U'. These pencils will form conics in the top and bottom plates, and these conics will be the sections of the quadric surface which will meet the cylinder in the required cubic.

Here is the method of working (see Figure 26.12). Through g we suppose any plane, which will meet the top and bottom plates in parallel lines f and f'. Through u there will be a corresponding plane of the relation ug, which will meet the top and bottom plates in parallel lines h and h'. These planes will have a common line which will be the join of the points fh and $f'h'$. This line will meet the cylinder in that generator of the cylinder which is contained in the plane of G determined by ff'; we have marked it P in the diagram and it is a point of the cubic. We note that the point fh of the top plate belongs to a conic determined by the UG relation, and once this conic is drawn, with its brother of the bottom plate, the cubic can be constructed very quickly and easily.

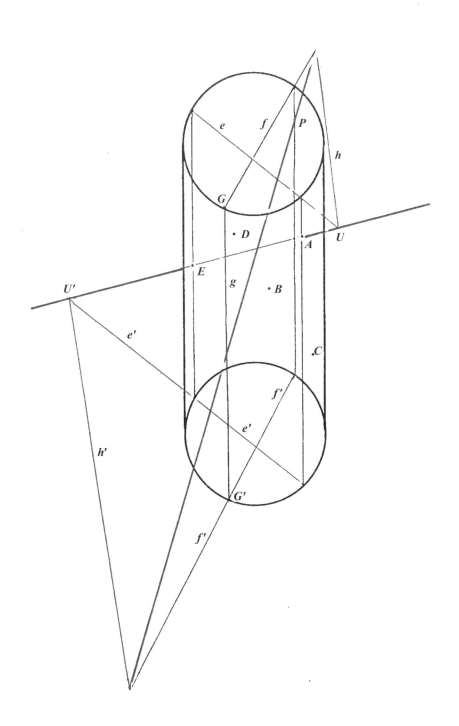

We have only one more problem to solve. How can we arrange the relations UG and $U'G'$ to ensure that the cubic will also pass through B, C and D?

Firstly, how can we ensure that the curve will go through B? We draw the vertical generator through B and this fixes for us the plane of g with which we must deal, meeting the top and bottom plates in f and f' (see Figure 26.13). The lines h and h' of U and U' must meet f and f' in points that will be collinear with B, remembering that h and h' must be parallel.

We reduce it to a plane problem (Figure 26.14). Lines f and f' are parallel; given U, U' and B, we have to find that pair of parallel lines through U and U' which shall meet f and f' respectively, in points X and X' which shall be collinear with B. Notice that the pairs of lines in U and U' being parallel must be projective, and they will meet f and f' in projective ranges. Consider the pair of U and U' which is parallel to f and f'. Clearly the point at infinity of f and f' is a self-corresponding point of the projectivity. Therefore the common lines of all pairs of corresponding points must be concurrent at some point. Clearly this point must lie somewhere on the line UU'. To find just where, draw any two parallel lines through U and U', to meet f and f' at N and N'. The line NN' meets the line UU' at the point of concurrency, O. Draw line

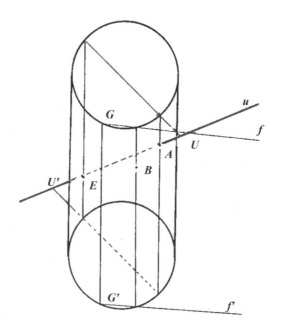

Figure 26.13

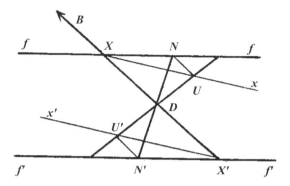

Figure 26.14

OB and let it meet *f* and *f'* at *X* and *X'*. The lines *UX* and *U'X'* are the parallel rays through *U* and *U'* which we are seeking.

We can thus find a pair of corresponding lines of *U* and *G* for each of the points *B*, *C* and *D*. These give three points for the conic in the top plane, which together with *G* and *U* completely determine it. The rest of the conic can then be constructed by the usual methods. The conic in the bottom plane appears by simply taking parallel pairs of rays through *U'* and *G'* to those through *U* and *G*.

Exercise 26b

Put five points on such a cylinder and carry out this construction. It is a very useful exercise and it is easier to do than to read about it!

26.9 A family of cubics on a hyperboloid

We can now extend the construction of Section 26.8 to draw a twisted cubic on the surface of a hyperboloid. For the sake of convenience we will again make the conics of our top and bottom plates circles with the same convention that the lower part of each circle is the part which is closer to our eyes. We can decide that the generator shall turn any definite amount, in one direction for the original set, and the other for the other, and the same amount for both sets, as they travel from one plate to the other. Let it be 90°. We then mark out our plates like in Figure 26.15.

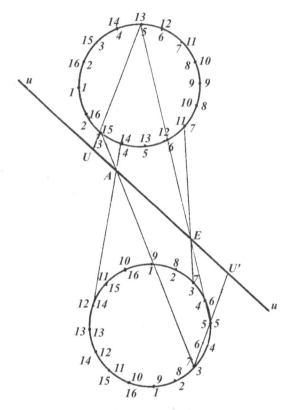

Figure 26.15

By joining any two outer numbers of the same value we obtain a
generator of what we will call the outer set, and we know that this will
be skew to every other such generator but will meet every generator of
the other set. We therefore join the 3s of the outer set and the 14s of the
inner and let them meet at *A*. Similarly we join 7–7 and 5–5 and let
them meet at *E*. *AE* is our line *u*, and we must find the points in which
it meets the top and bottom plates.

We notice that the outer generator through *A* and the inner one
through *E* have a plane in common and that *u* belongs to this plane.

Therefore if we join, in the top plate, 3 of the outer to 5 of the inner
we shall get the line in which this plane meets that plate, and this line
meets *u* in the point *U* which we are seeking. Notice that we could
equally have joined 7 outer to 14 inner of the top plate, and if we do
this we will find that it is concurrent with the other two lines at *U*.
Similarly we find *U'* for the bottom plate. We now choose any genera-

tor of the outer set, for our line g, and then carry out the construction of Section 26.8, using now only the inner generators.

We can now go on to a further very beautiful construction. We have been seeking to put a cubic through five given points on the quadric. In so doing we have arrived at a construction which involves having a conic on the top plane passing through U and G (and another on the bottom through U' and G'). If we are not particular as to which points on the quadric the cubic should go through we could put any conic we wish through U and G and go ahead to draw the cubic quickly and easily. The construction itself, using rays of the bottom plate parallel to those of the top, will automatically supply the appropriate conic for the bottom plate.

Now supposing that we have done this, we go on to put a further conic through U and G to meet the first one in two other points, say, S and T. It is clear from the very method of the construction that the points S S' and T T' will give the same points on the quadric for the new cubic as they did for the first one, and both cubics must, of course also pass through A and E. It is thus clear that by putting a family of conics through U, G, S and T we can construct a family of cubics lying on the quadric and passing through four common points. Line u is obviously bisecant to this family and thus also the line GG' must be — and thus every other outer generator. All inner generators will be unisecant.

By having a family of circles with two real common points on the top plate one can draw a family passing through two imaginary points and two real ones, A and E. By ensuring in the original choice of lines that the line u passes completely outside the hyperboloid (in which case U and U' can be chosen fairly freely) one can have a family of cubics on the quadric passing through four common imaginary points. This picture is a particularly beautiful and interesting one and is well worth doing. See Figure 26.16.

In Figure 26.16 the line UU' passes down the 'inside' of the hyperboloid and is just about asymptotic to it; that is, it touches the surface at infinity, or very nearly so. Thus the picture represents a family of twisted cubics on the hyperboloid, having four points in common, two imaginary, and a real co-incident pair at infinity. Were line UU' to be moved ever so little, so that it would just not touch the hyperboloid, the curves would approach to one another very closely as they approach infinity but would never actually touch one another.

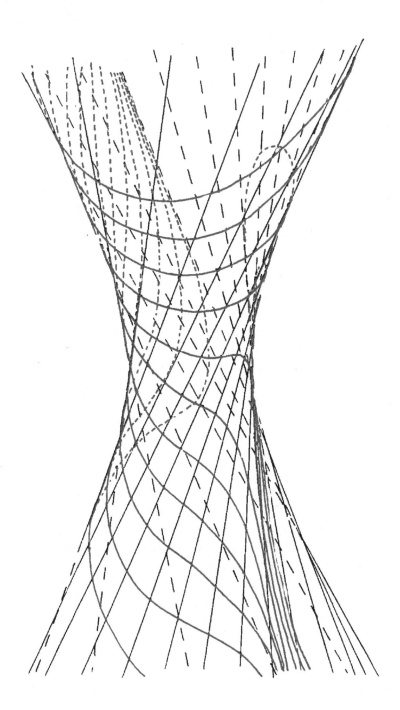

Figure 26.16

26.10 The twisted cubic seen from an eye-centre

We should now think of the dual aspect. Firstly it must be clear to us that when we see such a twisted cubic in space we are seeing it from the point-centre of our eye. All the lines joining our eye-centre to the cubic form a cone. Every plane of our eye-centre contains just three points of the cubic, therefore every plane of this centre contains just three lines of the cone; it is a cone of third order (Section 23.6).

Any plane section of this third order cone will be a third order plane curve. Thus when we look at a twisted cubic, and turn the model from one position to another, we are watching the various metamorphoses which plane cubics can undergo (with the exception of the irrational cubics, third order, sixth class which are not included in this series). Notice that when we have our eye in one of the osculating planes of the cubic we will see, looking along that plane, either an inflexion or a cusp.

However, if our eye-centre happens to be on one of the points of the curve, things are different. We know that every twisted cubic lies on ∞^2 quadrics, and among these are a single infinitude of cones, one for each point of the curve. These cones are all of second order and it was by the intersection of two of these that we made our first drawings of the twisted cubic in Section 26.3. Thus the cone of lines from any point of space to a twisted cubic is of third order, unless that point happens to be a point of the curve, in which case it is of second order.

26.11 The developable

If we dualize the things we have been doing we shall find ourselves dealing with curves of third class, made up of osculating planes. Each pair of neighbouring osculating planes will meet in a line, which will be a tangent line of the curve. All the tangent lines of the curve form what is called a *developable*.

Let us dualize the construction of Section 26.3. Two cones centered in points A and B have a common generator (line AB). We look for every non-skew pair of lines, one from each of A and B; this means we must look for every pair of lines which share a common plane, and this plane must obviously be a plane of the line AB. Such a pair of lines will have a common point, and this will be a point of the curve.

Instead of points A and B we will take planes α and β. Their common line is the line $\alpha\beta$ and touching this we must draw two curves of second class; that is, conics. Now we must seek a pair of lines (x and y) of these curves having a common point; this must obviously be on the common line of the two planes. These lines will have a common plane, and this will be an osculating plane of the developable. Now notice that momentarily line x is moving around its point of contact with the α-conic, and y is turning around its point of contact with the β-conic. Therefore it is clear that the osculating plane must be turning around the line which joins these two points of contact. This line therefore is a tangent line of the curve and a line of the developable. Hence the construction as in Figure 26.17.

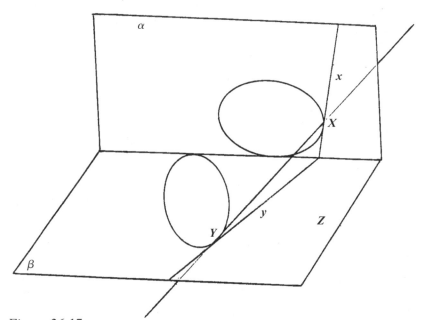

Figure 26.17

Take any line (tangent) of the α-conic, x, which touches the conic at X, and let it meet line $\alpha\beta$ at Z. Draw through Z line y tangent to the β-conic, to touch it at Y. Line XY is one of the lines of the developable. By going round the conics in this way the whole curve may be constructed.

Theoretically the conics may be placed in any relation to one another, provided only that they are both tangent to the common line of their two planes. However in practice, unless the relationships are well chosen the figure does not always come out very conveniently or beautifully. A model made in this way is very illuminating to study.

An obvious, and very simple, model to make is the following. We have two horizontal plates; therefore they meet on the line at infinity. Our conics must therefore each be a parabola. These curves can be of any size, and situated at random. All we now have to do is to join points whose tangents on the top and bottom plates are parallel. A beautiful cubic developable results.

Let us suppose that we have any point on any quadric. It will have one tangent plane, but there will be an infinitude of lines which will touch the quadric at this point, namely the pencil of lines in that tangent plane through its point of contact. All these are tangents to the quadric at this point. Any curve on the quadric passing through this point on the surface of the quadric, will have a tangent line at the point and this tangent is bound to be one of those lines; that is, it must lie in the tangent plane of the quadric at that point.

26.12 A developable osculating three given planes

Now we have seen that any six points of space will determine a twisted cubic, but these could be three double pairs. For instance, three points and the tangent lines through them will determine a twisted cubic (Figure 26.18).

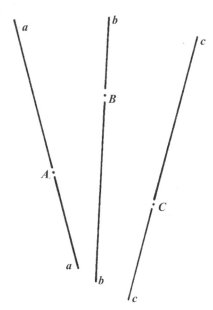

Figure 26.18

Let us determine a cone in the point *B*. Obviously it must contain the lines *BA*, *BC* and *b* (since the curve will pass through *B* along this line), and it must also contain as tangent planes *Ba* and *Bc*, because, by the above, *a* and *c* must both lie somewhere in a tangent plane of the cone. But three lines and two planes are enough to determine a cone. The cone in *C* is likewise determined by the lines *CA*, *CB* and *c*, and the planes *Ca* and *Cb*. These cones have a common generator in *BC–CB* and will determine the cubic we are seeking. Equally we could have used cones in *A* and *B*, or in *A* and *C*.

The intersections of these three pairs of cones give a particularly beautiful picture when they are dualized.

In doing this we shall, of course, come against the problem: given three lines in a plane, how do we construct a conic tangent to all three of them, and touching two of them at given points of contact? This is the dual of the problem set in Exercise 6g (page 79). It is suggested that the reader should put this book aside and work it out for himself before reading further. In case of failure, or lack of time, here is the solution.

Required to construct the conic touching lines *a* at *A*, *b'* at *B'*, and *x*. We must construct along *a* and *b'* suitable projective ranges. Such ranges will be swept out by the tangent *x* as it moves round the curve. Therefore the points in which *x* meets *a* and *b'*, *X* and *X'*, will be one corresponding pair. Now imagine a position of *x* very close to *a* (dotted in Figure 26.19). Clearly as it approaches *a* its meeting

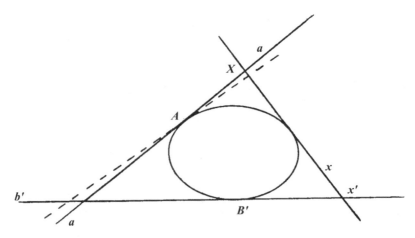

Figure 26.19

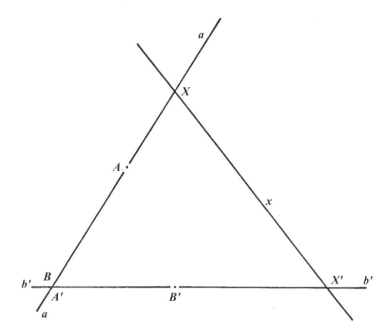

Figure 26.20

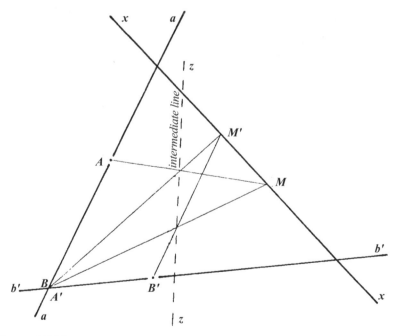

Figure 26.21

point with *a* will approach *A* and its meeting point with *b'* will approach the meet of *a* and *b'*. Therefore *A* of line *a* will correspond to *A'* of line *b*, as marked in Figure 26.20, and similarly *B'* of line *b'* will correspond to *B* of line *a* — also as marked.

Now we need two intermediate raying points *M* and *M'*, and an intermediate line *z*. We know (Section 6.7) that the line joining *M* to *M'* will meet the base lines, *a* and *b'*, in a pair of corresponding points of the projectivity, therefore it will be very convenient to put *M* and *M'* at any two positions along line *x*. Then *M'B'* and *MB* will meet on the intermediate line; so also will *M'A'* and *MA*. Thus we find the intermediate line and can go on to construct the desired conic (Figure 26.21).

Now we can start our dualizing of Figure 26.18. The points *A*, *B* and *C* will become three planes, *α*, *β* and *γ*; we will set them in the Cartesian relation, at right angles with one another. In each we put

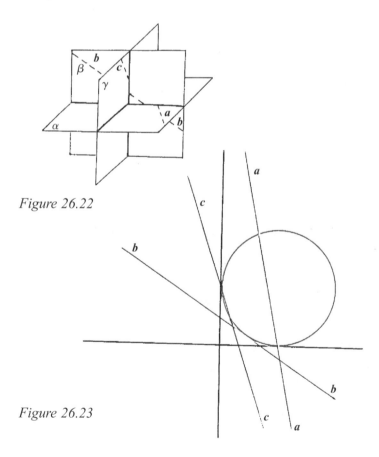

Figure 26.22

Figure 26.23

one line *a*, *b* and *c* respectively. The figure will work out very beautifully if we put the lines somewhat like Figure 26.22.

In order to save too much construction work we will use a circle in plane β. Put this in red, tangent to the vertical and horizontal axes, and draw line *b* tangent to it. Lines *a* and *c* will now be drawn, in their respective planes, passing through the points in which the red circle touches those planes. The conics in α and γ must touch the line in their own plane and must touch the rectangular axes at the points where these are met by the lines of the other planes. The conic in α turns out to be an ellipse, and that in γ to be a hyperbola.

One can now go on to construct a developable having the three Cartesian planes as osculating planes, and meeting each of them in conics. It is best to do the picture twice, showing only two of the planes being met in conics in any one picture, otherwise the page becomes overcrowded.

It would be most interesting to work out this figure with some other relationships of planes, conics and lines *a*, *b* and *c*. This particular relationship has been shown here partly to show the method, and partly for those with limited time, as it is known that this one shows all the most interesting parts of the model on the page and from a good viewpoint.

26.13 Curves of third order

Let us consider once again the view of the twisted cubic from one's eye-centre. From this centre is a single infinitude of lines, one going to each point of the cubic. Every plane through the eye-centre will contain just three points of the cubic, and therefore three lines of the cone which lies in the eye-centre. This cone is therefore one of third order.

Dualizing this, we 'view' a developable from any plane of space, and we find that every osculating plane of the curve will meet our plane in one line and these lines will lie in some curve. Every point of this plane will contain three lines (tangents) of this curve. Therefore our developable will meet every plane of space in a curve of third class.

Let us suppose we have a model with two plates, conveniently, but not necessarily, parallel. We put a curve of third class in each. Any osculating plane of the curve will meet the top and bottom plates in a pair of lines which will meet on the common line of the two plates; that is, a parallel pair of lines if the plates are parallel. These lines will be

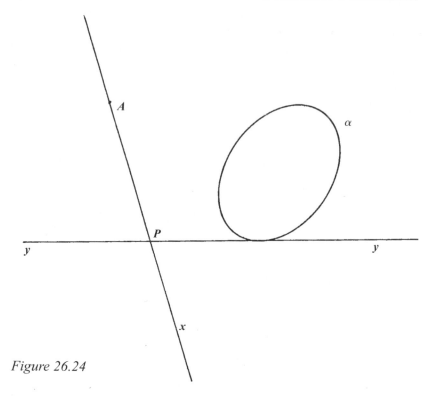

Figure 26.24

tangents of the curves and each line will be turning, momentarily, about its point of contact with its curve. It is clear that the osculating plane will be turning about the common line of these two points of contact. Thus the procedure: find a pair of points, in top and bottom plates, which have parallel tangents; join them and you have a line of the developable.

We are now faced with the problem of constructing third class curves. We refer to Section 20.1. There we studied the problem of establishing a one-to-one correspondence between single points on the one hand, and involutary pairs on the other, all being co-basal. We found that in such a case there are always just three *incident members* of the range; that is, just three of the single points will coincide with one or the other point of the corresponding involutary pair.

Now let us put the lines of a pencil into one-to-one correspondence with the tangents of a conic. The meets of all lines with their corresponding tangents will lie on some locus. What will it be?

Line x of pencil A corresponds in a one-to-one projectivity with tangent y of conic α (see Figure 26.24). Their meet P is a point of the curve whose locus we must now examine.

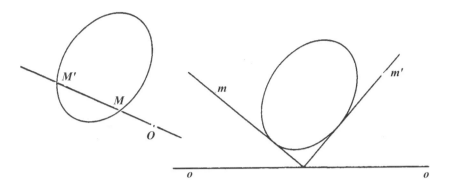

Figure 26.25

Firstly, let us refer back to Section 9.3. We remember that if we have any conic confronted by any point O, not on the conic, then the lines of that point meet the conic in pairs of points in involution. In Figure 26.25, M and M' are a pair in involution. In Figure 26.26 we see the dual of this. We confront a linewise conic with any line o, (not a tangent) and each point of that line carries two tangents which are a pair in involution on the conic, m and m' are a pair of tangents in involution.

Now we return to Figure 26.24. We put down any arbitrary line of the plane z. The tangents of the conic will sweep out a range of points along z, and so will the lines of A. These two ranges will be in a certain relationship. Consider any point Z, of z. It will carry two tangents to the conic, and these will be a pair in involution. They will therefore correspond to a pair of lines of A which must also be in involution, and which will meet z in a pair of points in involution. Therefore Z corresponds to a pair of points in involution. There will be just three incident members of the relationship, and these will be the points in which our locus meets line z. The locus is therefore a curve of third order.

This is an important construction and many beautiful diagrams can be done with it. There are many ways in which the relationship between the lines of the pencil in A and the tangents of the conic can be established. The obvious one is to let the lines of A meet some intermediate line in a range of points and to project these points onto the conic from some point of the conic. The tangents at these points will then clearly be projective with the lines of A.

Exercise 26c

Draw some curves with this construction. If you use a circle for the conic, and put A, the intermediate line and the raying point of the conic, into some symmetrical relationship you will get particularly satisfying pictures. Take every point of the circle, draw its tangent and mark its common point with its corresponding line of A. If you now do exactly the same thing with a selection of the imaginary points of the circle you will construct the imaginary part of the cubic — very lovely!

26.14 Curves of third class

The dual of the above would read like this: establish a one-to-one correspondence between the points of a conic and the points of a line. Join corresponding points and these lines will be tangents to a curve of third class. And this is the curve we are seeking.

Exercise 26d

Draw the dual of the above. We are going to let the points of line a be in a one-to-one relation with the points of circle α. We will use an intermediate point M and an intermediate line m, tangent to the circle, and place them symmetrically with regard to the other elements, as in Figure 26.27.

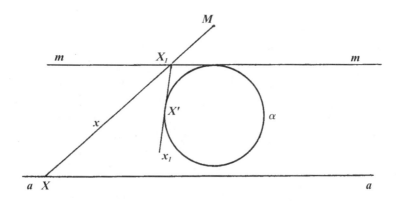

Figure 26.27

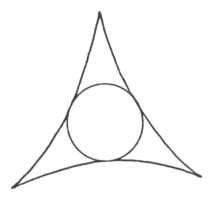

Figure 26.28

Now, any point X of line a corresponds to line x of M, and this corresponds to tangent x_1 of α, and this corresponds to point of contact X' of the circle. X and X' are corresponding points in the projectivity we are seeking and their join is a tangent to the third class curve. If you do this you will come to a three-cusped curve which, drawn pointwise, looks approximately like Figure 26.28.

26.15 To construct a model of a cubic developable

This curve is going to be specially useful and beautiful for a model if we can learn to draw it in its most regular form; that is, so that its three cusps lie exactly on the points of an equilateral triangle. How must M, m and a be disposed so that this result will come about?

It is clear from Figure 26.27, that that line of M which is tangent to the circle is also a tangent of the curve we are seeking. Obviously the circle will touch the curve at the midpoints of an equilateral triangle the apex of which will be point M, and the centroid of which will be the centre of the circle. (N.B. This is not the triangle on which the three cusps will lie).

We know, from elementary Euclidean geometry that the centroid of a triangle comes 1/3 of the way up the median. Therefore we see that M must be placed just twice the radius of the circle above O, its centre. We place m in the same position as before (Figure 26.29) and we now have to ask, 'Where must we place line a if the curve is to be regular?'

We notice that the tangents at the three cusps must clearly pass through the centre of the circle, so we draw ON at an angle of 120° with OM and let it meet the circle at N, Figure 26.30. We now have to

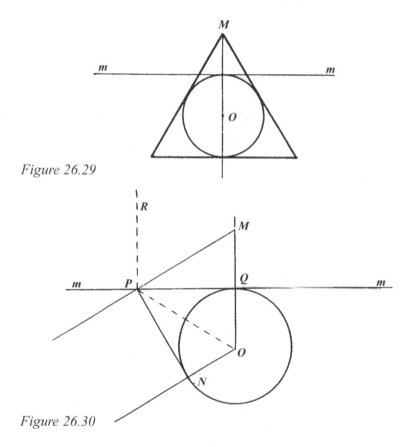

Figure 26.29

Figure 26.30

place line a in such a way that ON becomes a line of our curve. Draw the tangent at N to meet m at P. MP must meet ON on line a. Notice that we have right angles at N and Q, so that $PQON$ are concylic; therefore

$$\angle NPQ = 180° - \angle NOM = 60°$$

PO bisects this so $\angle OPQ = 30°$. Obviously triangles OPQ and MPQ are congruent, so $\angle QPM = 30°$, and $\angle NPM$ is a right angle. MP is parallel to ON and they meet on the line at infinity. We could do a similar thing for the third cusp and this shows clearly that if the curve is to be regular we must make line a the line at infinity.

Figure 26.30 gives us further valuable information. Let $\angle NOM$ be variable now, say θ. All tangents to the curve will still be parallel to MP but this will not, in general be parallel to ON. However $\angle NPQ$ will still be $180° - \theta$ and

$$\angle OPQ = \angle QPM = (180 - \theta)/2 = 90 - \theta/2$$

Draw *PR* perpendicular to *m*.

$$\angle RPM = 90 - (90 - \theta/2) = \theta/2$$

This inclination of *ON* to the perpendicular is always just twice that of the corresponding tangent to the curve. In other words, for any given tangent, the angular velocity with which it turns about *N* is half that with which *N* turns about *O*.

Let us consider such a tangent to our third class curve, through *N* (Figure 26.31).

Since *t* makes an angle with *RN* which is one half of angle *NOA*, *t* is the tangent to our third class curve. As *N* moves in the direction of its tangent, *NP*, *t* turns momentarily in a clockwise direction around *N*. We draw *NP* of unit length, and use it as a vector to represent the velocity of *N*. Through *P* we draw *PM* parallel to *t*, so that *NM* is perpendicular to *t*. The vector *NM* now represents the velocity with which *N* is moving in the direction of *NM*. But *t* is turning with half this angular velocity in a clockwise direction, so we produce *PM* to *Q*, so that

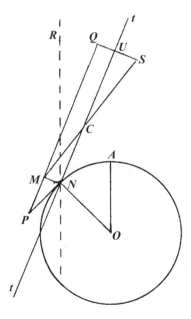

Figure 26.31

$MQ = 2NO$, and draw $QS = NP$ perpendicular to MQ. We let MS meet t at point C.

MN and QS are vectors representing momentary velocities. Notice that, by similar triangles, we can vary the size of vector NP as much as we like and the position of C remains unaltered. If in our imagination we allow MN and QS to become infinitesimally small we can see that C is the point around which the line will be turning. It is therefore the point of contact of this tangent with its third class curve.

Applying this method to Figure 26.30 we will see that OQM is the tangent to the third class curve passing through Q, and we find that the point of the cusp is at a point situated from O just three times the distance OQ. The circle circumscribing the curve has therefore a radius of $3OQ$.

We can now draw this curve for the top plate of a model using $\theta =$ every ten degrees. This will give a series of tangents whose inclinations to the perpendicular will be every five degrees. For the bottom plate we will turn the figure through 30° and we will find that $\theta + 60°$ will give a parallel tangent to θ on the top plate. We construct the points of contact of these tangents and join them up, point by point, and we shall have a third order developable meeting all horizontal planes, and the plane at infinity, in three real points.

The top and bottom plates, when superimposed on one another, look something like Figure 26.32. One line of the developable will go from A of the top plate to A' of the bottom, and another will go

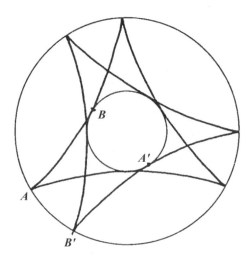

Figure 26.32

from B of the top to B' of the bottom. Clearly the curve itself will meet the top plate at A and the bottom at B', and by the symmetry of the case the line of the developable half way between these will touch the curve halfway between the top and bottom plates. The line for the top plate will be found at $\theta = 90°$, giving a tangent inclined to the perpendicular at 45°.

Next we must look for the point of contact of this tangent with its third class curve (see Figure 26.33). Through N we draw the tangent to the circle, $NP = 1$ unit length. Since the triangle PMN is isosceles and right-angled the vector $MP = 1/\sqrt{2}$. We draw QP parallel to NM and equal to $2r$, and QS right angles to it, equal to our unit vector, NP. SP meets MN at C.

Now by similar triangles

$$\frac{MC}{1/\sqrt{2}} = \frac{2r}{1} \quad \text{or } MC = r\sqrt{2}$$

But the true point of contact, C, will obviously be this distance away from N. Therefore if we make our circle centre O to have a radius of

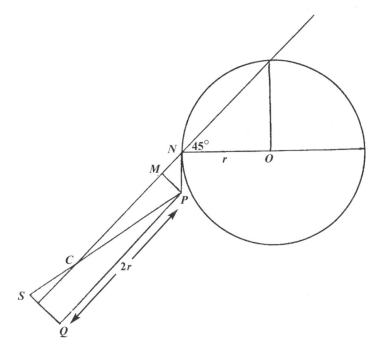

Figure 26.33

1, then the point of contact of this tangent with its third class curve will be $\sqrt{2}$ distant from N. By the cosine rule we can easily find that it is distant $\sqrt{5}$ from O.

The corresponding tangent for the bottom plate will have a value of θ equal to $90° + 60° = 150°$, giving a tangent inclined to the perpendicular 75°. Since however the bottom curve is turned through 30° this tangent will in fact be at 45° to the perpendicular as measured on the top plate making it parallel to the top plate tangent as required by the construction. By symmetry it is obvious that the point of contact for this tangent will also be $\sqrt{5}$ from O. We now put both these tangents on the one diagram (see Figure 26.34).

It is clear from elementary Euclidean considerations that the line CC' is at right angles to both NC and $N'C'$, and that therefore the triangle NCT is an isosceles right-angled one. Therefore $CT = \sqrt{2}$ and $NT = 2$. The line CC' therefore meets the circumscribing circle of the third class curve in two points U and T, 90° apart, (remembering that the radius of the circumscribing circle has already been found to be three times ON).

Seeing that the developable is clearly going to be symmetrical with respect to the top and bottom plates it is obvious that one of the hyperboloids on which it will lie will be one which will meet the top and bottom plates in the circumscribing circles of the third class curves, and

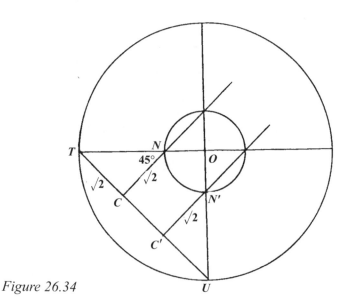

Figure 26.34

that this tangent line to the developable, *CC'*, will lie in the tangent plane of this hyperboloid at its waist. Therefore the hyperboloid will be constructed by joining points on the two, circumscribing circles which are 90° apart and these strings can be added to the model when the cubic has been constructed.

It is strongly recommended that the reader should make this model, after the general fashion described at the beginning of Section 24.1, and also the models described in Chapters 27 and 28.

27. More Developables

27.1 Curves enveloped by circles

An interesting, and sometimes useful way of constructing higher order curves is by making them from the movement of a circle — enveloping them with circles. The following is the well-known construction for a cardioid: we draw a circle and put any point K on it (Figure 27.1). Now we draw a selection of all the circles possible, having their centres on the original circle and passing through K. The result is a cardioid beautifully enveloped by circles. The large circle is the original one, and the smaller is one of the circles enveloping the cardioid.

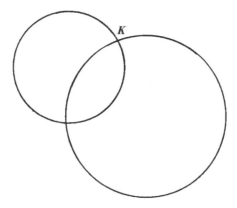

Figure 27.1

There is no need for point K to be on the circle. It can be outside it — in which case the curve, instead of coming to a cusp at K, passes through it with a little hump pointing towards the original circle. If K is inside the circle the curve will make an inward loop through K.

There is no need for the original curve to be a circle. It can be a hyperbola. If K is on the hyperbola a pear-shaped curve appears; if K is inside one of the branches of the hyperbola the pointed end of the pear becomes rounded, while if K is between the two branches of the hyperbola, the curve becomes a lemniscate. If the hyperbola is rectangular and K is at its centre, the lemniscate is the true Cassini one.

Exercise 27a

Draw a selection of such curves. Infinite variation is possible. Some of the irregular lemniscates are particularly beautiful. They are all beautiful and very easy to draw, and give quite a good experience of various types of higher order curves which are possible, although, of course, there are many other types which are not obtainable thus.

It would be going a long way outside the field of our present studies to show exactly the way these curves arise. They are intimately connected with the process of inversion which we have not yet dealt with, the centre of inversion being represented by point K. It is sufficient for us here to note that they are all curves of fourth order. When K is not on the original curve they are also of fourth class but, as can be verified by inspection, when K is on the original curve they become curves of fourth order and third class. The cardioid is one such, and is therefore a useful curve for making a cubic developable model.

Some very useful facts about the cardioid can be gathered easily from this method of construction.

Let us consider an enveloping circle of the cardioid, centred at A (Figure 27.2). Consider a neighbouring circle of the curve (one infinitesimally close to the circle centre A). The common point of these two circles will be the point round which they are, momentarily, turning; that is, the point of contact with the cardioid. Since the centres A and A' are neighbouring, their common line will be the tangent to the original circle at A. Now we know from ordinary geometry that all circles having their centres along this line, and passing through K must also pass through the reflection of K in the line; that is, through P where PK is the line bisected at right angles by the tangent through A. Thus we get a simple construction to show us the point of contact of any circle of the envelope with the cardioid. Draw the tangent to the original circle through the centre of the enveloping circle and find the reflection of K in this line. This is the point at which the enveloping circle touches the cardioid.

Now we let the angle KOA be θ.

$$\angle OAK = 90 - \theta/2$$
$$\angle KAX = \theta/2$$
$$\angle AKP = \angle APK = 90 - \theta/2$$

We let the tangent to the enveloping circle at P, which is clearly also the tangent to the cardioid at that point, meet OK at T.

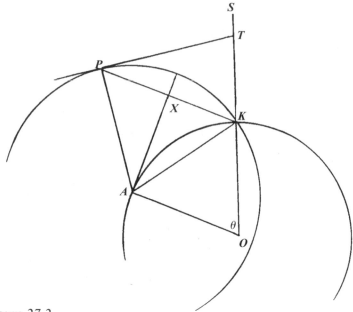

Figure 27.2

$$\angle KPT = 90 - \angle APK = \theta/2$$
$$\angle PKT = \theta$$
(note that KX is parallel to OA)
$$\angle PTS = \theta + \theta/2 = 3\theta/2$$

27.2 Another developable model

Thus we see that for every θ turned at O, the corresponding tangent to the cardioid turns through an angle of $3\theta/2$. So if we have a point P of the cardioid, constructed with an angle α at centre O, and if we now move 120° on and construct the point of contact given by an angle at O of $\alpha + 120$, we shall have a point whose tangent to the cardioid is turned 180° from that of the point P. In other words we shall have constructed two points whose tangents are parallel. The line joining them, from top to bottom plates will be a line of the developable.

We need not draw the enveloping circles as the points of contact are so easily gotten with the above construction; nor need we draw the tangents to the cardioid; we know that points obtained with lines from O separated by 120° will have parallel tangents.

Take your original circle, with K on it, and divide it into 15° intervals (Figure 27.3). At each of these intervals draw a tangent, drop a perpendicular on to it from *K*, produce it as far again on the other side, and this is a point of the cardioid.

Every point will have a tangent to the cardioid which is parallel to that of the point which is eight points on. If we join such points on our plane figure we get a linewise cubic (third order) which is in fact the view of the twisted cubic which we get by looking down on the model from above the top plate. We see that the closest point which the cubic gets to the centre of the circle, that is to the central axis of the model, is on the horizontal line of our diagram, *WW'*. Clearly this marks the

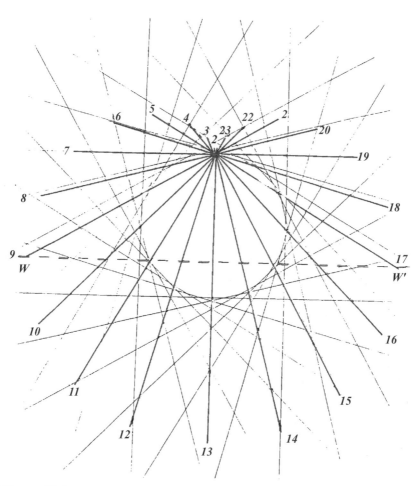

Figure 27.3

waist of the hyperboloid through the original circle on which the developable lies. We see that this line meets this circle in points which are 120° apart, so this gives the measure for constructing this hyperboloid.

To construct this model, one should bore holes in the same way for both the top and bottom transparent plates; that is, all the points, from 1 to 24 on the cardioid, and also every 15° round the circle. First one makes the hyperboloid by threading from each hole of the top circle to the hole of the bottom circle which is 120° away. Next, with a different coloured thread, one goes from hole No. 1 of the top cardioid to hole No. 9 of the bottom, then from No. 10 of the bottom to No. 2 of the top, then No. 3 of the top to No. 11 of the bottom, etc. One will find that one has constructed a tangent-wise twisted cubic (that is, a developable) which lies exactly and very beautifully on the hyperboloid.

This model is exceedingly instructive to make, and to study. In the first place it illustrates vividly the two different ways in which a line can move in space. The inner form, the hyperboloid, is made by a line moving skew to itself, and forming a true plastic surface. It contains ∞^2 points, and ∞^2 tangent planes, with a one-to-one correspondence between points and planes — to each point its unique tangent plane, and vice versa. The outer form, the developable, is what we might call a quasi-surface; it contains ∞^2 points, but only ∞^1 tangent planes, each line of points sharing just one tangent plane. Such a quasi-surface always has what is called a *cuspidal edge*; this edge hugs the hyperboloid all along its length, and it forms the actual curve of the twisted cubic.

Notice that by turning the model round in one's hand one can see the cubic in any of the forms which are illustrated in Figure 11.7 (page 133), from a humped cubic to a looped one. Intermediate between these is the cusped cubic. Whenever one is seeing it as a cusped cubic one has placed one's eye actually in one of the osculating planes of the curve. But if one places one's eye in the prolongation of one of the lines of the developable itself, one sees the curve in the form of a conic section; one is then looking along a generator of one of the second order cones on which the curve lies.

28. The Linear Congruence and the Tetrahedron

28.1 A family of quadric surfaces

We have seen that a linear congruence, that is, a congruence of first order, consists of ∞^2 lines chosen out of the ∞^4 possible lines of space, in such a way that each point of space contains just one line. The most obvious way to make one's choice is to choose all the lines which meet any given skew pair of lines.

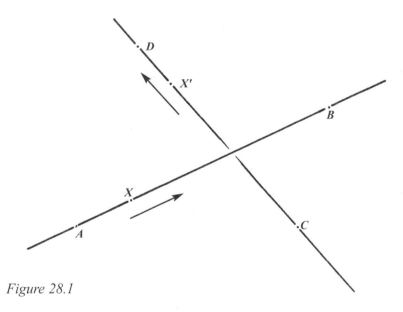

Figure 28.1

We have further seen that if we put growth measures on such a pair of lines, of equal cross ratios, we immediately determine a whole family (∞^1) of quadric surfaces. For if we have skew lines AB and CD (Figure 28.1), and we put growth measures of equal cross ratio between A and B, and between C and D, as double points, we have only to choose one other corresponding pair, say X and X', and then join all other points in order along the growth measures, and we will have a quadric. We remember that this quadric is determined by the three pairs of corresponding points, A–C, X–X', and B–D, or, what is the same thing, by the three skew lines joining these pairs (Section 5.3, page 64), a projectivity is determined by three corresponding pairs of entities). Further, that the form of the quadric is not affected by the cross ratio which has been chosen, but simply by the double points and the choice of $X X'$. The cross ratio determines how close together on the quadric the generators will lie.

Now let X and X' move in the direction of the arrows along the growth measures. The line XX' moves as a generator across the quadric surface, until in an infinite number of steps it would come to rest on the line BD. Clearly, AC and BD are generators of this quadric, as well as all the positions of XX'. Since AB and CD meet all positions of XX', as well as AC and BD, they also are generators of the quadric (of the other set).

If we had paired X with some other point in the first place, say X'', we should have obtained another quadric, but it would still have contained AC, BD, AB and CD as generators. There are ∞^1 points of CD with which we could have paired X, therefore we see that it is possible to have a family of quadrics having four common generators, provided that they are two of one set and two of the other. This is one of the spatial equivalents to a family of conics through four common points in the plane (Section 12.1).

28.2 Reguli within the tetrahedron

Now we must examine the tetrahedron. This is the simplest enclosed figure in space, thereby holding a similar position to the triangle in the plane. It is composed of four points, which we will call A, B, C and D, and opposite each point is just one plane which does not contain that point. These planes we will call α, β, γ and δ respectively. It contains six lines. Notice that, like the triangle, it is self-dual:

— four points, six lines, four planes
— through each point pass three planes and in each plane lie three points
— through each point pass three lines and in each plane lie three lines.

Notice that each line meets four of the others, and is skew to the fifth. Thus the tetrahedron naturally pairs its lines into three skew pairs, a fact of great importance.

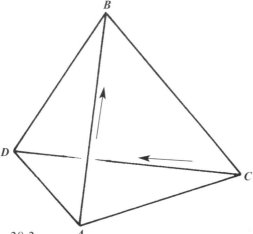

Figure 28.2

Now let us suppose that we have equal growth measures moving along *AB* and *CD* in the direction of the arrows (Figure 28.2).* We immediately have a family of quadrics containing *AB*, *CD*, *AC* and *BD* as common generators. Notice that the two generators passing through *A* are *AB* and *AC*, and that therefore, for all the quadrics of the family, plane *ABC* is the tangent plane at *A*; that is, it is the polar plane of *A* with respect to the quadric. Similarly the generators through *D* are *CD* and *BD*, so that the polar plane of *D* is plane *BCD*. The polar planes of these two points must meet on the polar line of their join. In other words the whole family of quadrics make lines *AD* and *BC* mutually polar. This is a general rule for any family of quadrics sharing four common lines of a tetrahedron, the family always makes the other, skew, pair mutually polar.

We will call the family which we have described above the *AB/CD* family, the order of the letters showing the direction in which the growth measures move. The family *BA/DC* would be identical with *AB/CD*, the movement of the generators just being assumed to be in the opposite direction across the surface.

We have shown that the regulus-family *AB/CD* contains *AC* and *BD* as generators and a moment's thought will show that the regulus-family *AC/BD* lies on the same quadric surfaces as the *AB/CD* family. These reguli are in fact the other sets of generators to those. There are

---

* As already mentioned at the beginning of this section, the motions on *AB* and *CD* are not essential for the following treatise. The only requirement is that there be a projectivity between these lines that maps *A* onto *C* and *B* onto *D*.

thus three families of quadrics lying in a tetrahedron, each one making one of the skew pairs of lines mutually polar.

Exercise 28a

There are 24 ways (permutations) to order four points. Verify that these group into *three* classes of eight permutations, each permutation within a class giving the same family of quadrics. These families can (and will) be represented by *AB/CD*, *AC/DB* and *AD/BC*.

We notice that each family of quadrics is made up from the lines of two congruences. For instance the *AB/CD* family is made up from the congruence of all the lines which meet *AB* and *CD*. In fact we could say that the lines of the congruence meeting *AB* and *CD* can be grouped into this family of reguli (they could also of course be grouped into many more such families by changing the double points of the growth measures along them, but these families would not be contained within the tetrahedron with which we are dealing).

But it is also made up from the congruence meeting *AC* and *BD*, these lines forming the conjugate reguli (other sets) of the quadrics.

Now if we change the direction of one of the arrows, and consider the family *BA/CD* = *AD/BC* we have a completely new family of quadrics. This family will include the lines *BC* and *AD*, which in the former set were mutually polar, and will make mutually polar the pair *AC* and *BD*, which were formerly included. But one set of reguli of this new family of quadrics (not the other) is made from the lines of the same congruence as formed the former (*AB/CD*) family. Thus we can say of the ∞^2 lines of any congruence, that they can be differently paired into *two* families of ∞^1 reguli, all lying completely within this tetrahedron, according to the way one groups them.

Now we must try to form an idea of what a family of quadrics, all sharing four common generators, looks like. It is hard to picture, and we shall only draw two quadrics of the family (see Figure 28.3).

These quadrics are of the family *AB/CD* = *AC/BD*. This family makes *AD* and *BC* mutually polar.

The top plate of the model is a horizontal plane and we see that this meets the model in a family of conics having double contact at both *A* and *D*. The tangents to these conics and *A* and *D* will lie in the common tangent planes to the hyperboloids, touching them at *A* and *D*. These planes clearly are, at *A*, *ABC*; and at *D*, *DBC*. Now imagine this horizontal plane tilted about the line *AD*. It will still obviously meet the

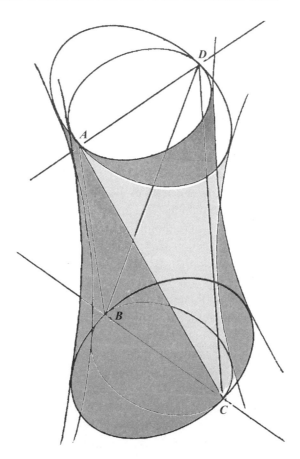

Figure 28.3

hyperboloids in a family of conics having double contact at *A* and *D*, whose common tangent lines will meet at the point at which the line *BC* meets the plane.

Consider a line moving along one quadric, starting at *AC* on the front of both hyperboloids, moving round as it proceeds towards *DB* at the back, always connecting *AB* with *CD*.* Its meeting point with the top plane would move along, say, the small ellipse, and its meeting point with the bottom one in the big ellipse.

The line *AC* is where the surfaces meet along the front, *BD* is their meeting line at the back.

---

* In order to get the ellipses of the figure it is supposed that the meeting point of the moving line with *AB* moves from *A* to *B* via infinity and the meeting point with *CD* moves between *C* and *D*; and then the other way round. If they both move between *A* and *B* and *C* and *D* respectively and afterwards both via infinity, the curves would be hyperbolae.

28.3 The mother form

Now we have to find what this will look like if the four points of the tetrahedron become imaginary, two by two in conjugate pairs. We let the line *AD* be a vertical, central axis, carrying the imaginary pair *A* and *D*, represented by an amplitude *XY*. We let the line *BC* be the line at infinity of the horizontal planes, carrying the points *I* and *J* in place of *B* and *C*. Our tetrahedron is now *ADIJ* and is all imaginary except for the two real lines. Next we take any plane of the line *IJ*; that is, any horizontal plane. How will it meet the model? Answer: in a family of conics having double contact at *I* and *J*, with their common tangents meeting at the point where the plane is met by the line *AD*.

We refer back to Section 16.4. There we pictured just such a family of conics, all having double contact at points *I'* and *J'*, with their common tangents meeting at a point *O*; and beside it we showed what the figure would become if the two points of double contact were to become the ordinary *I* and *J* on the infinite line. The result would be a set of concentric circles, centred at *O*. Thus we could make a model of a set of quadrics having four imaginary generators in common, *AI*, *AJ*, *DI* and *DJ*. It would be a system of hyperboloids meeting the top and bottom plates in concentric circles. If the distance between the two plates is equal to the amplitude of the imaginary pair *AD*, then all generators will stretch from points 90° apart on the two plates (see Section 24.11).

This model is so important, and so much comes from it that it is sometimes called the *mother form*. It comes in various kinds, and the reader is strongly advised to make at least two of them. They are very beautiful, and well worth the effort. Moreover it is not possible to illustrate them effectively, and for a real understanding it is needful to have them actually in one's hand.

The construction of the first one is illustrated in the upper part of Figure 28.4. One joins *S* to *S*, *T* to *T*, etc. from top to bottom plates, all the way round the circle, with coloured thread. Then using another colour one joins *W* to *W*, *X* to *X*, etc. all the way round that circle, and so on for all the circles, using a different colour for each circle.

The second model is shown in the lower part of the figure. One joins *S* to *S*, *T* to *T*, etc. all the way along the line. Then, using another colour one joins *W* to *W*, *X* to *X*, etc. all the way along the new line, and so on, for the whole model, using a different colour for each pair of lines, top and bottom.

In both models we see that we have actually strung up only a small

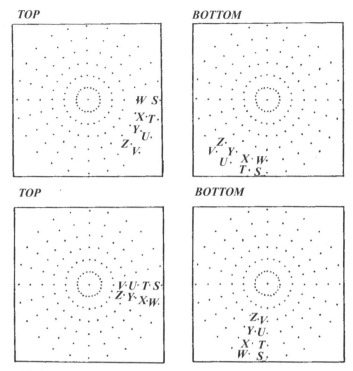

Figure 28.4

number of the possible lines of the set. In the whole model there would be one line to every point of the plane; that is, ∞^2, which is what we know there must be in a congruence.

The next thing to notice is that the lines we are using are identical for both models, yet the models have entirely different appearances,

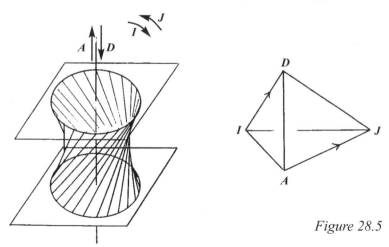

Figure 28.5

the first one being made of hyperboloids, and the second of paraboloids. This is simply due to the fact that the circling measure movement
along one of the original skew lines has become reversed; exactly the
same congruence is involved.

In Figure 28.5 we picture just one of the hyperboloids belonging to
the first model, and, on the right, the real analogue for this. Note that
as a point moves upwards along one of the generators of this hyperboloid, it is partaking of the movements of the imaginary points A and
J, whereas if it reverses its movement, and goes downward, it is partaking of the movements of I and D. This generator holds an imaginary
pair, one of which lies on the line AJ and the other on ID. Notice that
if we move to the next generator in a J-direction (that is, counter-clock-
wise) it is also lower on the hyperboloid; that is, it has moved in the D-
direction — hence the direction of the arrow on our real analogue — a
movement towards J is paired with a movement towards D. Looking
at the right-hand part of the figure it now becomes clear that when our
generator passes through A it also passes through I. In the left-hand
picture the imaginary line AI is a generator of the hyperboloid. (In fact,
these lines are common generators for the whole system of hyperboloids on this model.) This being so, and looking again at the real
analogue, since the two generators through A are AI and AJ it is clear
that the tangent plane at A is the plane AIJ, that is to say, this point and
plane are pole and polar with respect to the hyperboloid. Similarly, D
and the plane DIJ are pole and polar. Thus, we see from the real analogue that the common line of A and D must be polar to the common
line of planes AIJ and DIJ. AD and IJ are polar lines. When we look

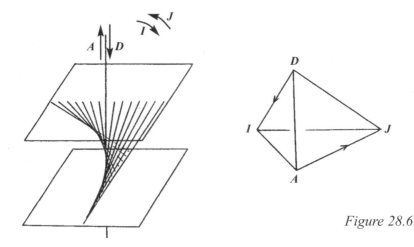

Figure 28.6

back to the left part of the figure we find confirmation of this — *AD*
and *IJ* (the horizontal line at infinity) are indeed polar to one another.

Looking now at Figure 28.6 we see the effect of reversing the movement along the line *DI*. We are still dealing with the same set of lines.
that is, the congruence meeting *AJ* and *DI*, but now it is clear that when
our generator passes through *A* it also passes through *D*; it becomes the
real line *AD*. Similarly it also becomes the real line *IJ*; that is, the horizontal line at infinity. And looking now at the left part of the figure we
see that these two real lines are indeed included among the generators
of the paraboloid.

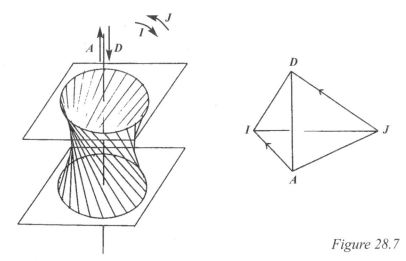

Figure 28.7

Looking back to our first model, we see that these lines imply
another congruence altogether — the lines of the conjugate reguli; that
is, the other set of generators which cross these on the hyperboloids.
Although this is another congruence it lies within the same tetrahedron, and in Figure 28.7 we picture it. Those lines which in the first
model were imaginary generators are now the generating skew pair for
the congruence *AI* and *DJ*, and what was formerly the generating skew
pair for the congruence has become an imaginary pair of generators for
the hyperboloid.

The second model also implied a new congruence, which is a set of
horizontal generators to the paraboloids. This is pictured in Figure
28.8. The generating skew pair for the congruence are now obviously
real. We are dealing with all the lines meeting the central axis *AD* and
the horizontal line at infinity. This one is not easy to make a model of,
but it can be quite easily imagined by looking at our second model.

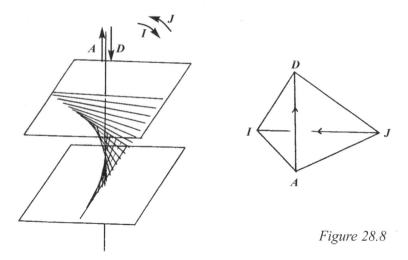

Figure 28.8

The same congruence as that shown in Figure 28.8 also contains another family of reguli, being paraboloids which coil in the opposite direction from the one in Figure 28.8. And these again imply another set of generators, being like those in Figure 28.6, but again with their surfaces coiling in the opposite direction. And these new generators are of the same congruence which we find in Figure 28.7. And with this we have made a complete round of the forms possible within this one tetrahedron.

We begin to see what a truly wonderful thing a tetrahedron is. It contains the possibility for three congruences (one for each skew pair of its lines) and each congruence contains two infinite families of reguli, of quite entrancing beauty. And these six infinite families are locked together in a circular inter-linkage either by having the same sets of generators lying on different quadric surfaces, or by having the same quadric surfaces carrying different sets of generators.

28.4 Skew imaginary lines

But there is more than this to be seen from these models. Earlier in this book the statement was made that these imaginary points and lines are, by their very nature, forever invisible. One can never do more than come to a concept of what they are, and are not like. But with the models of Figure 28.5 or Figure 28.6 in one's hand one is almost tempted to withdraw this statement. Let us look once again at Figure 28.5. The generators of the many hyperboloids which appear on this model represent for us the congruence of lines meeting the imaginary lines *ID* and *AJ*. Each of these generators contains a pair of imaginaries, one belonging to the line *AJ* and one to *ID*. Where are these imaginaries to be found? To answer this let us look at the left side of Figure 28.5, and imagine a horizontal line, lying on the bottom plate and going from the central axis, *AD*, to one of the generators. Now we know, from what has gone before, that if this lines rises up the central axis in one of the circling measures of the pair *AD* and simultaneously turns in a similar circling measure of the pair *IJ*, it will continue to meet the generator all along its length; and it will sweep out on that generator a circling measure of the pair of imaginary points in which that generator meets the lines *AJ* and *ID*. And now imagine all such circling measures on all the generators of all the reguli of model 28.5 and one will have a picture in one's mind of all the points of the imaginary pair of lines *AJ* and *ID*. In fact, a picture of a conjugate pair of skew imaginary lines!

But one can do more. Take the model of Figures 28.5 and 28.6, hold it level with one's eyes and look through it to a blank white screen. One will see hundreds of lines; they are all skew to one another, but they *appear* to intersect where in hundreds of places one line passes behind another. Now start the model slowly rotating about its central axis. Behold, all the apparent intersections will start to move, all in the same direction, and in perfect circling measure, more slowly in the middle part, and faster near the top and bottom plates. One is 'seeing' the circling measures of the points of the line *AJ*. Start the model rotating in the opposite direction, and all the intersections will move the other way; one will be 'seeing' the circling measures of the points of line *ID*.

In the plane, two lines always meet. In space, they may meet, or they may be skew. In the plane two conjugate imaginary lines always have a real common point — the centre of their circling pencils. In space it may be so, but it may also be otherwise; they may be conjugate skew imaginaries; then the thing they have in common is the model of Figure 28.5 or Figure 28.6, which is the set of common line-carriers for all their imaginary points. Anyone who has made and studied these models can well come to the conclusion that a pair of conjugate skew imaginary lines is one of the most beautiful things in the realm of thought.

29. Perspective

29.1 The picture plane and the field

Perspective is the science by which we depict a three dimensional *field* upon a two dimensional plane. Normally the spectator imagines between him and the field a vertical plane, placed at right angles to the direction of his vision when he looks straight ahead, and on this he makes a direct perspective of the objects of the field, with his eye as centre point. Thus all projective relationships of the field are retained on this plane (called hereafter the *picture plane*); for instance, three objects equally spaced in line in the field. no matter what the angle of the line with the picture plane, will come out on that plane as being harmonically spaced in line with the point of that line which is the projection of the point at infinity of the field-line. The reverse is not necessarily true; three objects in line on the picture plane are not, in general, representing collinear objects of the field.

In order to disentangle the complexities of this matter we shall need to consider three viewpoints. The first is the picture plane itself, and is what the spectator sees in the act of perspective. The second is the *side elevation* (Figure 29.1), and the third is the *plan* (Figure 29.2).

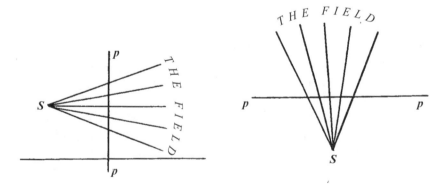

Figure 29.1. Side elevation.   *Figure 29.2. Plan*

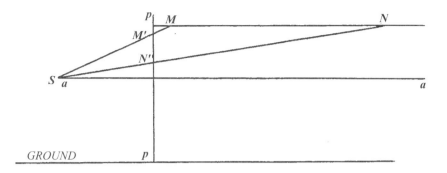

Figure 29.3

In the left Figure 29.1 we have gone round to the side, and the pic-ture plane, represented by line *p* meets through our eye. To the left we see the eye of the spectator, *S*, with his lines of sight passing through the picture plane to the objects of the field on the right, beyond. In the right figure (29.2) we see the whole thing from immediately above the picture plane, looking down on it. On the lower part of this figure we see the spectator looking to the objects of the field beyond the picture plane. We shall draw these figures, each one its own way up, as here, in order the better to distinguish between them.

Firstly, let us look at the side elevation (see Figure 29.3). A hori-zontal plane passing through the spectator's eye, *a*, in the figure, will clearly be seen as a horizontal line across the middle of the picture plane. What about a horizontal plane which passes, say a foot, above the spectator's head? Obviously an object, *M*, which is on this plane, and is very near to the picture plane will appear much higher above the central horizontal line of the picture plane, *M'*. Another object, *N*, also on this horizontal plane, but which is farther away will appear much closer to the central line, *N'*. By letting the object get farther and farther away, we see that the line at infinity of this other hori-zontal plane will appear as the central horizontal line of the picture plane. In fact, this line acts as the line at infinity for all horizontal planes. (A line which on the picture plane represents a line at infin-ity of the field will be called a *vanishing line*). This, the central hor-izontal line of the picture plane, is the vanishing line for all horizontal planes of the field.

The next thing we have to notice is that the particular perspective view which will result from any given field, depends on the position of the spectator with respect to the picture plane, in particular on his distance from it. Strictly speaking, if a perspective picture is made from one viewpoint (that is, from one distance from the picture plane) it can only be looked at, without distortion, also from that distance. Looked at from any other distance from the picture, there will be distortion, although this is not easily noticed in most cases.

We will suppose that the spectator is unit distance from the picture plane and we will now have a look at the plan (Figure 29.4).

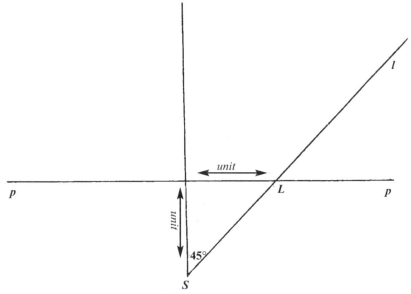

Figure 29.4

Let us suppose there is an object on the line at infinity of the horizontal plane, which is placed so that the spectator has to turn his eye through an angle of 45° from the centre of the plane, in order to see it. His line of sight is shown, *l*, on Figure 29.4, and it clearly meets the vanishing line at a point *L*, which is unit distance from the centre of the picture plane. Lines from *S* obviously make a circling measure along this vanishing line, and the points of this measure represent angular measure along the horizontal line at infinity. So we can say that angular measure along the line at infinity is shown on the picture plane as a circling measure having unity, (the distance of the spectator from the picture plane) as its amplitude.* In fact, we can say that the imaginary

---

* In this chapter there is a different definition of 'amplitude,' which equals half of the amplitude as defined in Section 13.3.

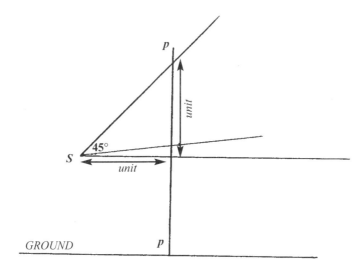

Figure 29.5

double points of this measure are the projection on to the picture plane of the horizontal *I* and *J*.

Now we look at the side elevation again, Figure 29.5. Suppose the horizontal plane now tips upwards at an angle of, say, 10° tilting about some line which is parallel to the picture plane. The spectator will now have to tilt his eye upwards through an angle of 10° in order to look along this plane to its line at infinity. Thus the vanishing line for this new plane will be a horizontal line, some way above the central line of the picture plane. Clearly all these vanishing lines will have their centres on the central vertical line of the picture plane and they will be in a circling measure.

Let us suppose that the spectator is looking upward at an angle of 45° to the vanishing line of a plane of the field which is tilted at that angle. His distance from the centre of that line is now, not 1, but $\sqrt{2}$ (by Pythagoras). We view this next in plan (Figure 29.6).

His distance from the vanishing line which he is now looking at is $\sqrt{2}$. If he now turns his eye to look at some object which is 45° to the right, measured along the line at infinity of this plane, this object will appear, not unit distance to the right along the vanishing line, but $\sqrt{2}$ to the right. Obviously the amplitude of the circling measure along this line vanishing line is $\sqrt{2}$. If the plane were to be tilted even further the amplitude of the circling meas-

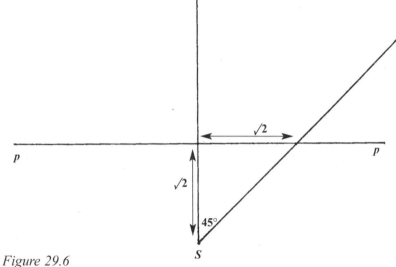

Figure 29.6

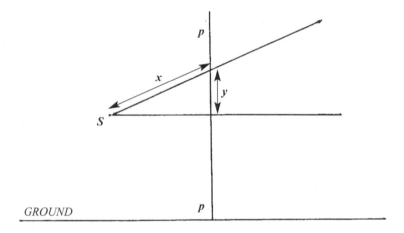

Figure 29.7

ure along the new vanishing line would be greater still. In fact, we
see that the amplitude is always equal to the distance of the spectator from the centre of the vanishing line in question.

We look at it again from the side elevation (Figure 29.7). We see
that if we let the distance of the vanishing line from the centre of the

picture plane be called y, and the distance of the spectator (that is, the amplitude of the corresponding circling measure) be called x,

$$x^2 = y^2 + 1$$

or putting it in the more usual form

$$x^2 - y^2 = 1$$

This is the equation of the curve which the amplitudes of the various vanishing lines which we have been considering, will follow across the picture plane. It is a rectangular hyperbola.

Now we have seen this before. We refer to Section 18.1 where we have exactly this picture. It is the picture of the amplitudes of the imaginary points of an imaginary circle, with real conjugate radius of 1. Here on our picture plane there appears the projection of the *absolute imaginary circle* in the plane at infinity, that circle which controls all the metrical relationships of three-dimensional Euclidean space. By its means we are enabled to move freely and with assurance among the laws of perspective.

29.3 To draw a cube

In order to show the method we will construct a perspective picture of a cube (Figure 29.8). We will view it looking down on it from an angle of depression of 20°. This is equivalent to looking straight ahead and tilting the cube through 20° towards us.

Therefore we will make the top plane of the cube lie on a plane which has its vanishing line 20° up. We draw AB to meet this vanishing line at, say, 30° to the left. We next draw a line from A to the 60° point on the right. This will be at right angles to AB, but we do not yet know just where to put C so that $AC=AB$. To find this we join B to the 75° point on the left, making an angle of 45° with AB, and this will meet the other line at C. Now we join C to the 30° on the left and B to the 60° on the right and these meet at D. We have made a true square on our 20° tilted plane. All lines perpendicular to this plane will pass through the point which is polar to it with respect to the absolute imaginary circle at infinity (Chapter 17). In our picture plane this means that any line which is at right angles to this 20° plane, must pass through the pole of the 20° vanishing line, with respect to the imagi-

nary circle which controls the perspective. This pole is found at the 70°
point below. So we join *A* and *C* to this point.

Now these two vertical lines lie in the front vertical plane of the
cube, and, being parallel, they meet at the 70° point below. This then
is one of the points at infinity of this plane. But the line *AC* also lies
in this plane, so the 60° point to the right, on the 20° line is also a
point at infinity of the vertical plane.

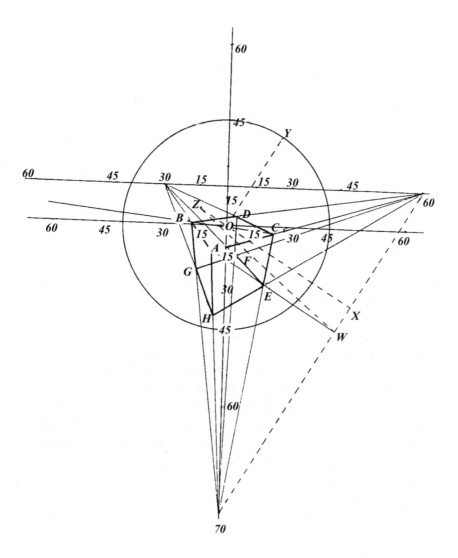

Figure 29.8

It follows that if we join this 60° point to the 70° point below we have drawn the vanishing line of the front vertical plane (dotted line in figure).

Now we need to construct the circling measure along this line. Drop a perpendicular from O, the centre of the circle, onto this line to meet it at X, and draw OY at right angles to this, to meet the real conjugate circle at Y. We know (Chapter 20) that the distance XY is the amplitude of the circling measure which the imaginary circle makes on this line, and this is also the distance from which we can project this circling measure (by equi-angular lines) from a point outside the line. So we produce XO to Z so that $XZ = XY$. If all has gone well we will now find that the 60° point on the right of the 20° line, and the 70° point below, will subtend exactly 90° at Z. They must do if the construction we are using is to make sense, and it lies within the laws of the imaginary circle that they will do. Next we bisect the angle which these two points make at Z, and let this bisector meet the vanishing line at W. W is now the 45° point between the 60° point and the 70° point. Draw AW and this will meet the vertical line through C (that is, the line joining C to the 70° point below) at E. We have now ensured that CE is not only perpendicular to AC but also equal to it. Through E draw a line to the 60° point on the right, and this produced will meet the vertical line through A at H.

Now the cube can be quickly finished. Join H to the 30° point on the left and this line will meet the vertical line through B at G. G to the 60° point on the right and E to the 30° point on the left will meet at F. You will now find that DF passes exactly through the 70° point below.

This cube appears at first sight to be rather distorted, by being drawn in an exaggerated perspective. In fact, it is exactly right when viewed from the right distance, which is the amplitude of the circling measure across the central line; that is, the radius of the conjugate circle of the imaginary circle which controls the perspective. Draw the picture gradually closer to your eye, and you will see how the cube gradually 'straightens out,' until, when it is the radius of the circle distant, it looks exactly right.

29.4 The stereoscopic view

If we now wish to make a stereoscopic view we must return to the plan-view of our spectator.

The distance between one's eyes is about 6 to 7 cm. Imagine the

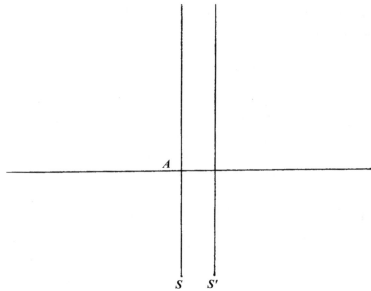

Figure 29.9

spectator changing his viewpoint by this amount, say to the right. He moves from *S* to *S'* (Figure 29.9). A point at infinity will project on to the picture plane in a point which will also move the same distance to the right. But if we imagine the front point of our cube, *A*, for instance, to be right on the picture plane, the projection of this will not move at all. All points at intermediate distances will move appropriate amounts across the picture plane.

Thus our method is clear. We move all the infinite elements of our picture plane a chosen distance to the right, and keep the closest point, in this case *A*, fixed. Then we just repeat the construction. In actual fact it is found much more convenient to keep all the infinite elements fixed and to move the point *A* by the chosen amount. This comes to exactly the same thing as long as we remember that when we move *A* to the right we are constructing the picture for the left eye. If our stereoscope is the old-fashioned reflecting type the mirrors will again interchange left and right after this!

The first displacement of point *A* must be a horizontal one, naturally, seeing that the line connecting our eyes is normally horizontal. By the fundamental projective laws it follows that all other points in the second picture will also be horizontally displaced. And indeed the construction would not work if it were otherwise. It is then a very good check of the correctness of one's work to see that all displacements are in fact truly horizontal. But this can be more than a check; it is a most useful adjunct to our construction methods. Having drawn the line through A' to the 30° point on the left we wish to know where the new point B' appears on it. Answer — by the method of horizontal displacement: B' must come at exactly the same height above the central horizontal lines as B. Often this method will be found quicker, and to be more exact, than carrying out the full construction. However where the lines are nearly horizontal themselves it becomes inaccurate. Having drawn our cube twice — once for each eye — like this, we can easily put the octahedron within it, and thus produce a truly three-dimensional effect of what we started with in this book, Figure 2.1 (page 21). Photograph these, and put them into an ordinary stereo-scope, and you will see the whole thing in three dimensions. It is a very fascinating exercise, to take many of the three-dimensional figures and to treat them like this. An easy one to be going on with would be Figure 3.2 (page 36). And there are many others.

Do not forget when you come to various problems in true perspective, that the absolute imaginary circle at infinity is the key to all your solutions.

Index

The Vortex of Life

Nature's Patterns in Space and Time

Lawrence Edwards

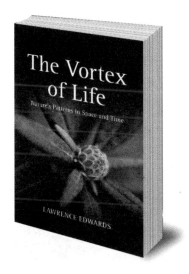

When *The Vortex of Life* was first published in 1993, Lawrence Edwards's pioneering work on bud shapes had already attracted the attention of many scientists around the world. In the book, Edwards gave a fuller account of his research, widening it to include the forms of plants, embryos and organs such as the heart.

His work suggests that there are universal laws, not yet fully understood, which guide an organism's growth into predetermined patterns. His work has profound implications for those working in genetics and stem-cell research.

This is a revised edition of the classic work edited by Graham Calderwood after Edwards' death in 2004.

florisbooks.co.uk

Mathematics in Nature, Space and Time

John Blackwood

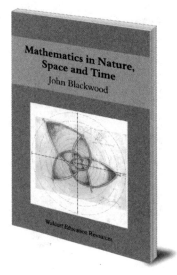

A teachers' book for maths covering 'Mathematics in Nature', 'Pythagoras and Numbers', 'Platonic Solids' and 'Rhythm and Cycles'. Includes full colour illustrations and diagrams throughout.

A resource for Steiner-Waldorf teachers for maths for Class 7 (age 12-13) and Class 8 (age 13-14).

This book is a combined edition of *Mathematics Around Us* (for Class 7) and *Mathematics in Space and Time* (for Class 8).

florisbooks.co.uk

Drawing Geometry
A Primer of Basic Forms for Artists, Designers and Architects

Jon Allen

This invaluable source book for students and professionals provides step-by-step instructions for constructing two-dimensional geometric shapes.

Making Geometry
Exploring Three-Dimensional Forms

Jon Allen

This unique book demonstrates how to make models of all the Platonic and Archimedean solids, as well as several other polyhedra and stellated forms.

florisbooks.co.uk

The Hidden Geometry of Flowers

Living Rhythms, Form and Number

Keith Critchlow

'a book of inspiration and insight'
– DR RUPERT SHELDRAKE

A beautiful and original book in which renowned thinker and geometrist Keith Critchlow focuses on flowers as examples of symmetry and geometry. Fully illustrated with hand-drawn geometric patterns.

florisbooks.co.uk